GRAPH THEORY - ADVANCED ALGORITHMS AND APPLICATIONS

Edited by **Beril Sirmacek**

Graph Theory - Advanced Algorithms and Applications
http://dx.doi.org/10.5772/65595
Edited by Beril Sirmacek

Contributors

P. Titus, A. P. Santhakumaran, Victor Chulaevsky, Daniel C. Mayer, Auparajita Krishnaa, Reji Kumar Karunakaran, Manuel Shibu, E N Satheesh, Francisco Casanova-Del-Angel, Izabela Kutschenreiter-Praszkiewicz, Pierre Mazzega, Claire Lajaunie, Etienne Fieux

Published by InTech
Janeza Trdine 9, 51000 Rijeka, Croatia

Publishing Process Manager Dajana Pemac
Technical Editor SPi Global
Cover InTech Design team

First published February, 2018
Printed in Croatia
Legal deposit, Croatia: National and University Library in Zagreb

Additional hard copies can be obtained from orders@intechopen.com

Graph Theory - Advanced Algorithms and Applications, Edited by Beril Sirmacek
p. cm.
Print ISBN 978-953-51-3772-6
Online ISBN 978-953-51-3773-3

For EU product safety concerns:
IN TECH d.o.o., Prolaz Marije Krucifikse Kozulić 3, 51000 Rijeka, Croatia,
info@intechopen.com or visit our website at intechopen.com.

Meet the editor

Beril Sirmacek holds a PhD degree in Electronics Engineering. During her PhD study, she has visited and worked in the most active remote sensing data processing laboratories around Europe (such as the remote sensing laboratory at University of Trento in Italy, Technical University of Munich in Germany, and French research institution INRIA in France). Following her PhD research, she started working as a research fellow at the German Aeronautics and Space Research Centre (DLR) in 2009, and besides that, she started to go through a three-year long habilitation research and study in the Institute of Geoinformatics at the University of Osnabrück in 2011. During her time in Germany, she has designed and gave lectures in German universities (Technical University of Munich, University of Osnabrück, University of Konstanz, and University of Erlangen-Nuremberg). In 2012, she started to work for TU Delft where she contributed by developing software for two FP7 European projects. Since 2015, she has been developing a cloud computing framework for supporting augmented reality mobile applications on mobile devices. Her shared company farmAR is now working with the European Space Agency at SBIC Noordwijk.

Contents

Preface

"Mathematics is the most beautiful and most powerful creation of the human spirit." ~ Stefan Banach

Most of the structures (physical, theoretical, social, algorithmic, etc.) in real or virtual scenarios can be represented by means of a diagram, which consists of a set of points (graph nodes) and their relationships between them (graph connections). This fact underlines the importance of graph theory knowledge in so many areas of our lives.

The links between graph theory and other branches of mathematics are definitely so strong. However, the importance of graph theory knowledge in developing algorithm such as classification, deep learning, and database management is becoming much more obvious. With the increasing amount and complexity of data, more scientists are going toward graph theory-based developments.

Graph theory does not only appear in books as a mathematical theory, but it actually helps scientists, program developers, decision makers, and engineers every day as well.

During my own PhD study, I was designing novel mathematical approaches for remotely sensed image processing. My aim was to make algorithms work fully automatically in order to make maps from images. That required developing very complex recognition systems, which can distinguish different urban objects (such as buildings, cars, parking lots, and trees) even when their appearances are very different in different areas and under different illumination conditions. My algorithms were becoming heavier, more difficult to debug, and more difficult to run on a regular computer. When I was really in trouble with designing more functional algorithms, I was introduced to graph theory. I was amazed by seeing the power of graph theory in simplifying difficult segmentation and matching problems in object recognition. Graph theory enabled to solve very complex problems easily by applying simple and intelligent matrix operations. That gave me the opportunity of writing more robust and faster algorithms. Seeing beauty and benefits of mathematics, I found myself diving into graph theory books even though I was not from mathematics field as an applied scientist.

Later, I had a lot of benefits from learning graph theory when I wanted to develop my skills in machine learning and deep learning methods. After some years of academic work, I have built my own businesses. Maybe it will not surprise you; indeed, I had a lot of benefits of graph theory knowledge when I was building on my R&D and business plans and product design cycles as well.

This book is prepared as a combination of the manuscripts submitted by respected mathematicians and scientists around the world. As an editor, I truly enjoyed reading each manuscript. Not only did the methods and explanations help me to understand theories better, but also I found it joyful to think about how to apply these theories in applied science developments. I believe that the readers will enjoy to read the book from the beginning to the end at once. However, the book can also be used as a reference guide in order to turn back to it in later years of research, implementation, or study.

I have to indicate this book assumes that the reader has a basic knowledge about graph theory. The very basics of the theory and terms are not explained at the beginner level. Therefore, if the reader is interested to read a chapter and does not have any pre-knowledge in this field, I would suggest to refer to other books that explain the very basics of the theory.

I would like to thank InTech publication family for their constant support and dedication to bring high-quality results.

I would like to thank all the authors for their kind collaboration by carefully following the review comments and applying the required changes. I also appreciate their efforts for making the visuals clear and understandable in order to make their writings more introductive to readers.

I would like to dedicate this book to all manuscript writers and the reader who is kindly considering this book. I wish it serves to the theoretical and applied science branches. If the readers would like to reach me or check my activities, they can follow my personal website at http://www.berilsirmacek.com

With my best regards,

Beril Sirmacek
create4D / CEO, owner
Rotterdam, Netherlands

Section 1

Graph Theory in Applications

Chapter 1

Spreading Information in Complex Networks: An Overview and Some Modified Methods

Reji Kumar Karunakaran, Shibu Manuel and
Edamana Narayanan Satheesh

Additional information is available at the end of the chapter

http://dx.doi.org/10.5772/intechopen.69204

Abstract

The knowledge of node's ability and importance in spreading information in a complex network is important for developing efficient methods either to decelerate spreading in the case of diseases or to accelerate spreading in the case of information flow, which would benefit the whole population. Some systems are highly affected by a small fraction of influential nodes. Number of fast and efficient spreaders in a network is much less compared to the number of ordinary members. Information about the influential spreaders is significant in the planning for the control of propagation of critical pieces of information in a social or information network. Identifying important members who act as the fastest and efficient spreaders is the focal theme of a large number of research papers. Researchers have identified approximately 10 different methods for this purpose. Degree centrality, closeness centrality, betweenness centrality, k-core decomposition, mixed degree decomposition, improved k-shell decomposition, etc., are some of these methods. In this expository article, we review all previous works done in the field of identifying potential spreaders in a network.

Keywords: social networks, information diffusion, node centrality, m-ranking, k-shell decomposition, improved k-shell decomposition, weighted k-shell decomposition, directed networks, degree centrality, closeness centrality, betweenness centrality

1. Introduction

Identifying and analyzing various kinds of network have become an important theme in the frontiers of research for the past 50 years. It is an emerging area which demands research activities of interdisciplinary and collaborative nature. Network research spreads over a variety of fields such as Mathematics, Physics, Chemistry, Computer Science, Biology and Social Sciences. Recently, network research has proved its importance by establishing itself as a new domain of research. It brings together and coordinates activities and research findings of researchers from other fields. Based on the nature and type of subject area and research, the entire field is divided into many subfields. Social networks, biological networks, chemical networks, physical networks, information networks, etc., are some of the most important ones.

Networks are being used for representing, analyzing and explaining complex systems. Complex systems have structural and behavioral characteristics. Both characteristics influence each other. Structural properties enable us to classify networks as random networks, scale-free networks, small-world networks and so on. Networks have been almost completely studied on the basis of its structure. Current scenario in the field is to establish relationship between the structural properties and the behavior of the network. When we speak of behavior, the behavior of the members and the behavior of the entire network come into picture. Behavior of the entire network is either the average of the behavior of the members or the totality of the behavior of the members.

An important area in the behavioral study on a network is the study of flow of information among the members. Information in this context of study is very general, which can be interpreted as spread of virus either in human contact networks [1, 2] or in a computer network [3, 4]. In some context, it is a spread of person-specific information in ego networks [5], or rumor that spreads in social networks [6, 7]. Irrespective of the nature and content of the information, there are some important factors which greatly affect the whole process of information diffusion in a social network. Structure of the underlying network is one factor. It is a global property. Ego networks and neighborhood of each actor in a network are important properties, which are classified as local properties. Rumor spread in social networks has got special attention as a kind of information spread. Rumor is unverified and instrumentally relevant information, the origin of which is uncertain and usually spread by word of mouth. Gossip and urban legend are also forms of information which spread in a social network by the way of social contacts. Character of the actors who receive a piece of information and transmit to other is a third factor, which would also greatly affect the information diffusion process.

In this chapter, we concentrate on some aspects of network-specific properties that influence transmission of information. Role of actors in a network varies according to their relative importance in the network and difference in their character. Characteristic difference and its effect are far from the scope of this chapter. Details of contacts in a network give us very fruitful information about the transmission process. For example, if the network is completely connected, a piece of information can be received by any other in a single step. In a disconnected network, the chance of not receiving the information by some actor cannot be completely removed. If the structure of the network is like a tree, average time taken for the information to reach the nodes depends on the place where it is generated. If the source of information has connection with all the remaining actors, the spread can end in one step. Networks can be modeled as graphs in Mathematics in which nodes represent interacting elements and edges represent interaction among the interacting elements.

Importance of nodes in the transmission of information is a matter of great concern in literature of information diffusion in networks [8–10]. As mentioned in the previous paragraph, some nodes that have large number of connections are highly important in the sense that any piece of information in their hands may spread faster than that in the hands of others. In other words, degree of nodes in a network is a measure of the importance of nodes. Centrality is a measure of a network that characterizes the importance of nodes. In other words, centrality measures are useful to analyze how "central" an individual node in a network. It is a function that assigns a real number value to each vertex in a network. Based on the value of this function, we can rank the nodes in the network. This measure in some context is used to study flow of something in a network. In some other context, it is a measure of cohesion among actors [11]. A centrality measure which is meant for some purpose may not give the right conclusion in a different context. In the following section, we review some important centrality measures along with an analysis of its ability to rank nodes. Each centrality measure is developed either as a generalization of an existing measure or as a new method which rectifies the shortcoming of an existing methods. We also propose a new measure of centrality which considers the importance of all nodes and edges present in a network while calculating the function value of each node. This measure can uniquely rank each different node in a network in a different class. If two nodes are equally important in the network topology, they fall in the same class. Thus, this method is seems to be more powerful than all other methods.

Clear knowledge of powerful or influential nodes in a network is valuable for many reasons. Certain nodes play a key role in the propagation of information or in the spreading of disease. If we know the role of the nodes in dissemination of information, we can manage to control the speed of information transfer through the network. This technique has great applications in the control of disease and in blocking the diffusion of annoying information such as rumors, negative behaviors, spreading of virus and so on. It can also be used in the campaigning of government, political parties, NGOs, public agencies and even in advertisements.

In general, networks are dynamic. Dynamic networks change over time. In a dynamic network, some nodes which are very important in some point of time may lose its status at some later point of time. Some nodes which are initially inactive may become very important through the interactions with other members in the network. Centrality measures are classified as local metrics, global metrics and hybrid metrics. There are many centrality methods which can be used to rank spreaders such as degree, k-shell, mixed degree decomposition, neighborhood coreness, extended neighborhood coreness and so on.

2. Comparison of important centrality measures

In this section, we discuss important centrality measures and relative importance. We also mention their drawbacks as evident from various contexts.

2.1. Degree centrality

In 1978, Freeman introduced the degree centrality [12]. He asserted that the degree of a focal node is the number of adjacencies in a network, that is, the number of nodes that the focal node is connected to. This is a basic centrality, and it is often used as a first step in the study of a network. The degree measure can be formulated as:

$$k_i = C_D(i) = \sum_{j}^{N} x_{ij} \tag{1}$$

where i is the focal node and j represents all other nodes. N is the total number of nodes. X is the adjacency matrix, in which $x_{ij} = 1$, if node i and j are connected and 0, otherwise.

The degree of a node is defined as the number of other nodes connected directly to a given node. In degree decomposition, we rank the nodes of a network according to the descending order of degrees of nodes. Nodes with highest degree are ranked 1, nodes with next higher degree as rank 2 and so on. This ranking method is known as degree decomposition. The complexity of this method is $O(n)$, where n is the number of nodes. It is a local metric. This is the first centrality measure that appeared in literature.

As an illustration, a network is given in the **Figure 1** and degree of all vertices is calculated. There are six vertices of degree one. These vertices are labeled as 1, 2, 3, 7, 9 and 15. The vertex labeled 14 has degree two. The vertices 5, 6 and 13 have degree three, and the vertices 4, 10, 11 and 12 have degree four. The vertex 8 has degree six. In the sense of degree centrality, the vertex 8 is the most powerful in the network and the vertices 4, 10, 11 and 12 occupy the next lower position. In this class, there are four members. A close look in the network will reveal the fact that the status of the vertices 11 and 12 is same. But 10 is more powerful than 11 and 12, because it is connected to the most powerful member in the network (vertex 8). But 4 in that group is much weaker than 10, 11 and 12. Clearly, this ranking scheme lacks accuracy.

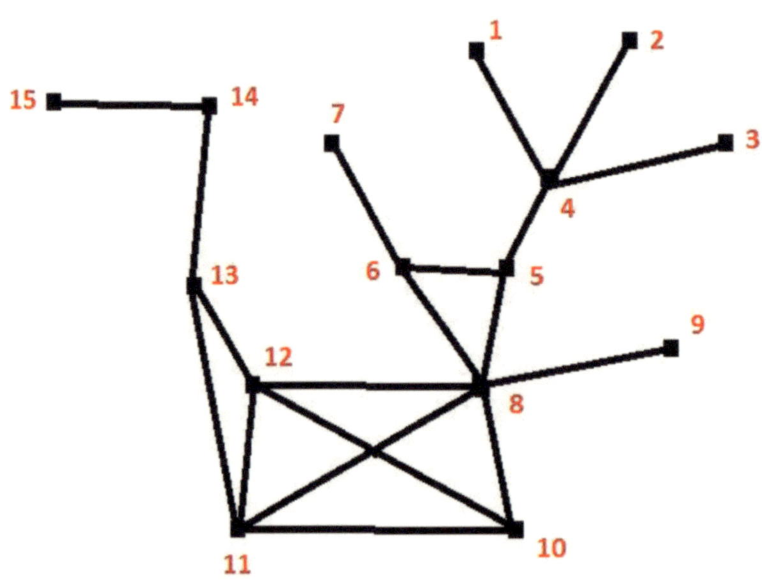

Figure 1. An example network containing 15 nodes.

In the case of weighted networks, degree centrality can be extended as sum of weights. This measure has been labeled as node strength and formulated as follows:

$$S_i = C_D^W(i) = \sum_j^N w_{ij} \tag{2}$$

where W is the weighted adjacency matrix, in which w_{ij} is greater than 0 if node i is connected to node j, and the value represents the weight of the link between i and j. This is equal to the usual degree centrality if the network is binary.

In both case, node strength is a rough measure, as it only takes into consideration a node's total connections with other nodes in the network. Clearly, degree and strength are both indicators of the level of connections of a node to other nodes of the network. It is important to incorporate both these measures for studying the new centrality of a node.

In Ref. [12], a new centrality measure is proposed. It is the average of the number of nodes connected to a focal node and the strength of that node. It is calculated using the following formula.

$$C_D^{w\alpha}(i) = k_i^{(1-\alpha)} s_i^{\alpha} \tag{3}$$

In the formula, α is a parameter that can set between 0 and 1. If α is near 0, the degree is more important, and if it is near 1, then node strength is more important.

Directed networks add complexity to degree centrality as two additional aspects of a node's involvement are to be incorporated in the calculations. The activity of a node can be quantified by the number of ties that originate from a node, denoted by K^{out}, and the number of ties that are directed toward a node, denoted by K^{in}. For a weighted network, S^{out} and S^{in} can be defined as the total weight attached to the outgoing and incoming ties, respectively. Taking into account the weights of the directed ties, the previous formula is modified in [12] to assess a node's importance in a directed network:

$$C_{D-out}^{w\alpha}(i) = k_i^{(1-\alpha)-out} s_i^{\alpha-out} \tag{4}$$

and

$$C_{D-in}^{w\alpha}(i) = k_i^{(1-\alpha)-in} s_i^{\alpha-in} \tag{5}$$

The value of α in these equations is defined as above.

2.2. Closeness centrality

The closeness centrality was also proposed by Freeman [13]. It is based on the length of the shortest paths among nodes in a network. In a binary network, the shortest path is found by minimizing the intermediary nodes, and its length is defined as the minimum number of ties linking the two nodes, directly or indirectly. We define the shortest distance between the vertices i and j as

$$d(i,j) = \min(x_{ih} + \ldots + x_{hj}). \tag{6}$$

Here h stands for intermediary nodes on a path between nodes i and j. The basic assumption in shortest distance is that the intermediary nodes increase the cost of the interaction and the higher number of intermediary nodes increases the time taken for interaction between nodes. Again intermediary nodes are powerful third parties and can distort information or delay information passage between nodes. In unweighted networks, the shortest path of two nodes is through the smallest number of intermediary nodes.

Closeness centrality relies on the length of the paths from a node to all other nodes. Closeness centrality is defined as:

$$C_c(i) = \left[\sum_j^N d(i,j)\right]^{-1} \tag{7}$$

We know that diseases are more likely to be transferred from one person to another person if they have frequent interaction. Frequency of interaction can be quantified and used as weights of links. In weighted networks, distance between two nodes i and j is defined as

$$d^w(i,j) = \min\left(\frac{1}{w_{ih}} + \ldots + \frac{1}{w_{hj}}\right). \tag{8}$$

where h is intermediary nodes and w_{ih} denotes its weights.

Here, closeness centrality is defined as

$$C_c(i) = \left[\sum_j^N d^w(i,j)\right]^{-1}. \tag{9}$$

In Ref. [12], Opsahl et al. propose a new closeness centrality by introducing the tuning parameter α, in finding the least costly path. This ensures that the both the weights of the ties and the number of intermediary nodes are considered in the identification of length of the path. Thus, length of the path is defined as

$$d^{w\alpha}(i,j) = \min\left(\frac{1}{(w_{ih})^\alpha} + \ldots + \frac{1}{(w_{hj})^\alpha}\right). \tag{10}$$

In the formula, α is a tuning parameter. If $\alpha = 0$, the proposed measure produces the same output as in the case of unweighted network. When $\alpha = 1$, this method gives the same result as the method proposed by Freeman. A value $\alpha < 1$ assigns the path with greatest number of intermediary nodes, the longest distance. Conversely, $\alpha > 1$, the impact of additional intermediary nodes is relatively unimportant compared to the strength of the ties and paths with more intermediaries.

2.3. Betweenness centrality

Betweenness centrality is measured for each vertex in a network. It quantifies the number of times each node appeared as a bridge along the shortest path between any two other nodes. This measure was also proposed by Freeman in 1958 [12]. Betweenness centrality for unweighted networks is defined as

$$C_B(i) = \sum_{j \neq i \neq k} \frac{g_{jk}(i)}{g_{jk}}, \tag{11}$$

where g_{jk} is the number of shortest paths between nodes j and k and $g_{jk}(i)$ is the number of shortest paths between nodes j and k and passing through node i. This quantity is added over all pairs of vertices in the network. In the case of weighted networks, betweenness centrality is defined as

$$C_{B^{w\alpha}}(i) = \sum_{j \neq i \neq k} \frac{g_{jk^{w\alpha}}(i)}{g_{jk^{w\alpha}}} \tag{12}$$

This can also be generalized to directed networks. The identification of the shortest path, and their lengths, in directed networks is similar to the process of undirected networks with an additional constraint. A path from one node to another node cannot be treated as connection in the reverse direction. Thus, a directed path is said to exist only if all edges in the path are directed in the same direction. As a result, the values of betweenness centrality calculated for a directed graph may significantly vary from that of the underlying undirected graph.

2.4. K-core decomposition

In 2010, Garas et al. put forward a fast node ranking method called k-shell decomposition [14] for large networks. The k-shell or k-core decomposition method partitions a network into substructures. This method assigns an index k_s to each node, which is the rank of the node in the network, according to its importance. Nodes with high values of the k_s are located at the center or core of the network, and nodes with low values of k_s lie in the periphery of the network. This way, the network is described by a layered structure, exposing the hierarchy of its nodes.

Now, we present a calculation of k-shell for nodes in the network of **Figure 1**. First, we removes recursively from the network all nodes of the network with degree 1 and we assign the integer value $k_s = 1$ to them. This procedure is repeated iteratively until there are only nodes with degree $d \geq 2$ in the network. Subsequently, we removes all nodes with degree $d = 2$. Again, this procedure is repeated iteratively until there are only nodes with degree $d \geq 3$ left in the network and assign to them the integer value $k_s = 2$ and so on. This procedure is applied until all nodes of the network have been assigned to one of the k-shells. Thus, we partition all nodes of the network into different shells with an integer value. This is how the original k-shell decomposition method works.

From the above network, we can assign $k_s = 1$ to the vertices 1, 2, 3, 4, 7, 9, 14 and 15 in the first step. Then, we can assign $k_s = 2$ to the vertices 5, 6 and 13. Finally, the vertices 8, 10, 11 and 12 get the value $k_s = 3$. These vertices are the most important in the network according to the assumptions of k-shell decomposition method. There are three categories of vertices. Each contains more than one vertex, which are of different nature. Hence, we arrive at the conclusion that this method is far from attaining the aim of the method. Computational complexity associated with the k-shell method is $O(n)$. It is a global metric. As a means to overcome this inefficiency of the k-shell decomposition method, a modified method called mixed degree decomposition method was proposed.

2.5. Mixed degree decomposition method

It was in 2013 that Zeng and Zhang proposed a modification of k-shell method called the mixed degree decomposition (MDD) [8]. The method is described as follows. The k-shell method is a dynamical network decomposition procedure in which the residual degree (number of links connected to the remaining nodes) of nodes is updated in each step while all the information of the removed nodes is dropped. In mixed degree decomposition method, both residual degree and exhausted degree (number of links connected to the removed nodes) of the nodes are recorded and the decomposition is based on both of them. For node i, exhausted degree is denoted by $k_i^{(e)}$ and residual degree is denoted by $k_i^{(r)}$. In each iteration, the nodes are removed according to the mixed degree defined by

$$k_i^{(m)} = k^{(r)} + \lambda k_i^{(e)} \tag{13}$$

where λ is a tunable parameter between 0 and 1. The detailed decomposition is described below.

1. Initially, $k^{(m)}$ of each node is equal to 0, since there is no removed node in the network.
2. Remove all nodes with least degree, denoted by M and assign them to the M-shell.
3. Update $k^{(m)}$ of all the remaining nodes by $k_i^{(m)} = k^{(r)} + \lambda k_i^{(e)}$. Then, remove all nodes with $k^{(m)}$ smaller than or equal to M and assign them to the M-shell also. This step is recursively carried on until $k^{(m)}$ of all remaining nodes are larger than M.
4. Repeat steps 2 and 3 as M value increases until all nodes in the network have been assigned to one of the shells.

When $\lambda = 1$, this MDD method coincides with the degree centrality method, and when $\lambda = 0$, this method coincides with the usual k-shell decomposition. The MDD method is no longer integer since $k^{(m)}$ can be decimal when we take λ between 0 and 1. The following simple example (**Figure 2**) illustrates the procedure of MDD method.

If a virus originates from a node with large exhausted degree, not only it has the same probability as the other nodes in the same shell to infect the nodes in the higher shells, but also it has a bigger branch of nodes in the lower shells to infect, so that this virus will end up covering much more

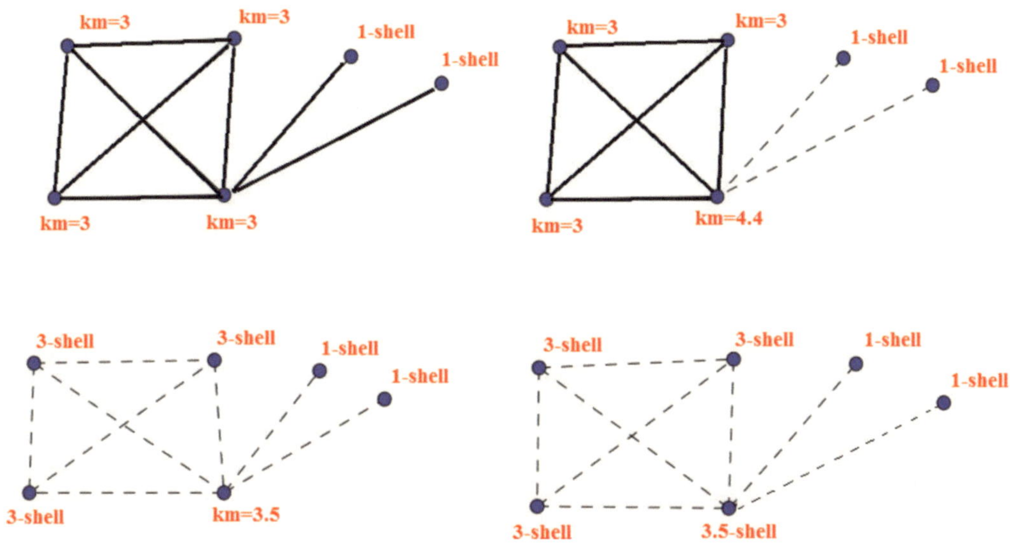

Figure 2. An illustration of MDD method.

nodes at the end. Thus, the information of the exhausted degree cannot be overlooked when ranking the nodes. The frequency of appearance of different nodes in MDD method is higher than in k-shell decomposition. The k-shell has limited number of ranks, and the number of nodes in each rank is quite high, which means that the node differences are not well distinguished in k-shell decomposition. In MDD method, the nodes more differently ranked than k-shell decomposition. So this method is preferred over the k-shell decomposition method. Liu et al. proposed an improvement on the k-shell decomposition in the year 2013. It is discussed below.

2.6. Improved k-shell decomposition method

Improved k-shell decomposition method [9] is calculated in terms of the distance from a target node to the network core, the spreading influence of the nodes within the same k-shell values could be distinguished. The formula for improved k-shell decomposition is that $\theta(i/k_s) = (k_s^{\max} - k_s + 1)\sum_{j\in J} d_{ij}$, where $k_s^{\max}$ is the largest k-core value of the network, d_{ij} is the shortest distance from the node i to the node j, and j is the set of nodes whose k-shell value is the maximum. Using this $\theta(i/k_s)$ value, we can rank the nodes of the network. Calculated values of θ for the network in **Figure 1** are as given below. $\theta(15) = 52$, $\theta(1) = 45, \theta(2) = 45, \theta(3) = 45$, $\theta(4) = 33$, $\theta(7) = 33$, $\theta(14) = 30$, $\theta(9) = 21$, $\theta(5) = 14$, $\theta(6) = 14$, $\theta(13) = 12$, $\theta(8) = 3$, $\theta(10) = 3$, $\theta(11) = 3$, $\theta(12) = 3$. Here, nodes of the network are divided into eight different shells using the new method.

In Ref. [15], Young Deng et al. proposed a refinement for the k-shell decomposition for unweighted network.

Here, the weight of the edge connection between i and j is defined as $w_{ij} = k_i + k_j$, where k_i and k_j are the degrees of node i and j, respectively. Next, for each node i, we calculate the weighted degree k_i^w using the following measure

$$k_i^w = \alpha k_i + (1 - \alpha)\sum_{j \in \Gamma_i} w_{ij}, \tag{14}$$

where Γ_i is the set of neighboring nodes of i, and α is a positive tuning parameter between 0 and 1 that can be set according to the requirement and data. If this parameter α is near 1, then the degree of nodes is taken as favorable. If $\alpha = 1$, this method becomes the usual k-shell decomposition. If α is close to 0, then the high edge weight is taken as favorable. Since the weighted degree may be no longer an integer, the weighted degree is approximated to the nearest integer. After this, the pruning routine is same as the k-shell decomposition method but is based on the weighted degree k_i^w. Using this method, we can partition the node set of the network into more shells.

Consider the network given in **Figure 3**.

Using the usual k-shell decomposition, the nodes are portioned as follows. The vertices 1, 2, 3, 4, 7, 9, 14 and 15 have $k_s = 1$. Vertices 5, 6 and 13 have $k_s = 2$. Vertices 8, 10, 11 and 12 have $k_s = 3$.

By the weighted k-shell decomposition (Wks) described above, the node partitions obtained are given in **Table 1**. Thus, the new method partitions set of nodes into seven shells, which indicate this method is more efficient (**Table 1**).

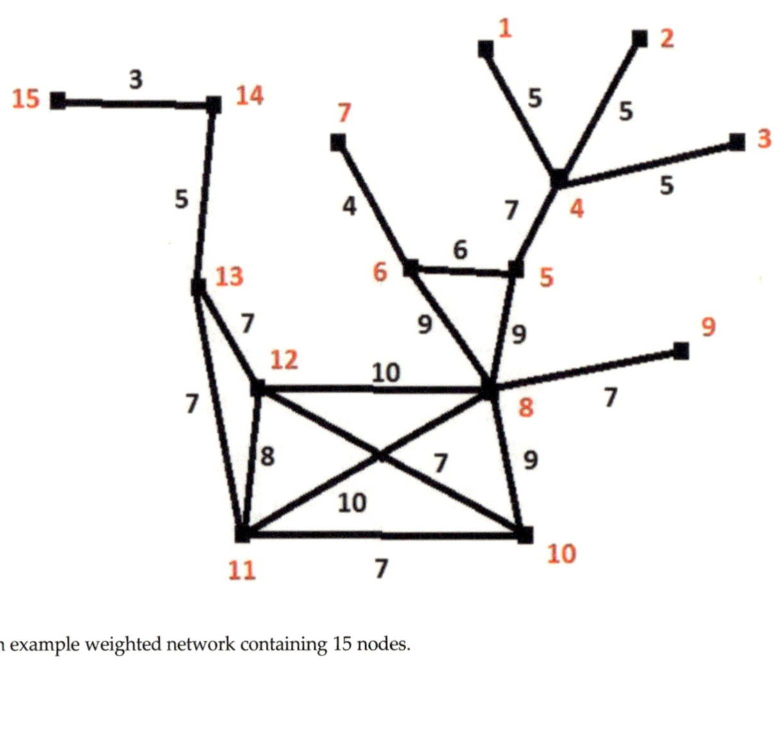

Figure 3. An example weighted network containing 15 nodes.

Sl. no.	Value of w_{k-s}	Vertex labels
1	1	7, 15
2	2	1, 2, 3
3	3	9, 14
4	4	4, 5, 6, 13
5	5	10
6	6	11, 12
7	7	8

Table 1. Ranks of nodes in the network given in **Figure 3** by Wks.

An improved version of the weighted decomposition method was suggested by Reji Kumar et al. [16]. The calculation of the rank of the nodes is based on the following procedure. The total of weight k_i^w of the vertices are calculated to obtain the value $T(G)$. So $T(G) = q\sum_{i=1}^{n} k_{i^w}$. Here n is the number of nodes. The right-hand side is multiplied by q to make $T(G)$ an integer. Next, the vertex 1 is deleted to obtain a new graph $G_1 = G - \{1\}$.. Subsequently, we find $T(G_1)$. Then, we delete the vertex 2 to obtain $G_2 = G - \{2\}$ and find $T(G_2)$ exactly as before. This procedure is repeated to obtain $T(G)$ values of all vertices. Finally, we find the value $T(G) - T(G_i)$ for the vertex i. The vertices are ranked on the basis of these values.

2.7. Decomposition using k-shell iteration factor

Very recently, Wang and Zhao put forward a method for fast ranking of influential nodes [17] in complex networks. He also improves the k-shell decomposition method using iteration factor. The method is as follows. Suppose node n_i is a node in the network and its k-shell value is k. In the k-shell decomposition, the total iteration number is m and n_i is removed in the nth iteration of the k-shell decomposition, where $1 \leq n \leq m$. Let δ_{ni} denotes the k-shell iteration factor of node n_i which is defined as follows;

$$\delta_{n_i} = k\left(1 + \frac{n}{m}\right). \tag{15}$$

The k-shell iteration factor of each node can thus be obtained with this formula. For example, for the above unweighted figure for node 3, $k_s = 1$, $n = 1$ and $m = 2$; therefore, $\delta_{n_3} = 1.5$. For node 4, $n = 2$, and $m = 2$, and $\delta_{n_4} = 2$. Here, nodes 3 and 4 are in the same k-shell, and this method distinguishes them with different δ value. Thus, if we rank the nodes of the network with the calculated δ value, this method will give a refinement of k-shell decomposition.

2.8. Weighted k-shell decomposition

One major limitation of most centrality measures, including k-core decomposition method, is that they are designed for unweighted networks. However, in practice, real-world networks are weighted, and their weights carry very valuable information about the strength or pace of flow of information through the edges. This information is highly critical in the study of transmission of information. In a weighted network, every node has two underlying attributes, their degree and weight of the edges incident with that node. Since weights are associated with networks links, the nodes weights are calculated as the sum over all link weights incident with a particular node. It may happen in network that a node with high degree can have small weight and vice versa. There are situations in which weights play important role. In such networks weights are related to some measured property. For example, in trade flow, capital flow, etc., nodes with high weight can usually be an important player. Thus, in such systems, the presence of nodes with high degree and relatively low weights may vary from the results obtained by methods that are based only on the degree of nodes. Under this approach, one completely neglects the weights and performs the analysis on unweighted network. In the second approach, consider only links and their weights and neglects degree.

Garas et al. gave the following method for weighted networks [14], called weighted k-shell decomposition method (W_k-shell). This method applies the same pruning method as in k-shell decomposition. Here, we consider both degree of a node and weights of its links. We also assign for each node a weighted degree k'. The weighted degree of a node is defined as

$$k_i' = \left[k_i^{\alpha} \left(\sum_j^{k_i} w_{ij} \right)^{\beta} \right]^{\frac{1}{\alpha+\beta}}, \tag{16}$$

where k_i is the degree of the node i, and $\sum_j^{k_i} w_{ij}$ is the sum of all link weights. α and β are two parameters. If $\alpha = \beta = 1$, the weight and degree equally get equal consideration in the calculation. If $\alpha = \beta = \frac{1}{2}$, then the above equation becomes

$$k_i^1 = \sqrt{k_i \sum_j^{k_i} w_{ij}}. \tag{17}$$

In the case of unweighted networks, where $w_{ij} = 1$, the weighted degree is equivalent to the usual node degree and we use the same usual k-shell decomposition method. In general, the weighted method is able to split further the cores obtained by the unweighted method and help us find the most central of the central nodes. Thus, it is reasonable to assume that the weighted k-shell partitioning method provides us with a more accurate node ranking for representing the node's spreading power.

In a recent article, Reji Kumar et al. [18] proposed a method in which degree and weights of all vertices and edges get relative representation in a network. This method is named as m-ranking method. This method is discussed in the subsequent section.

2.9. The m-ranking of nodes

Total power is calculated for each node in the m-ranking method. Total power of a node i is defined by the formula,

$$T(i) = \left\{ \alpha \left[d_i^{(0)} + \frac{1}{\beta} \sum d_i^{(1)} + \frac{1}{\beta^2} \sum d_i^{(2)} + \ldots \right] + (1-\alpha) \left[\sum W_i^{(1)} + \sum W_i^{(2)} + \ldots \right] \right\}. \quad (18)$$

In the formula, $\alpha = \frac{p}{q}$ is a parameter between 0 and 1 and $\beta > 1$ is another parameter. It is good practice to choose β an integer close to the average degree of the nodes of the graph. The first series contains at most $D + 1$ terms, and the second series contains at most D terms, where D is the diameter of the graph. Here, $d_i^{(0)}$ is the degree of the node i, and $\sum d_i^{(j)}$ is the sum of the degrees of the nodes of the graph at a distance j away from node i. Similarly, $\sum W_i^{(0)}$ is the sum of the weights of the incident edges of the node i and $\sum W_i^{(j)}$, is the sum of the weights of all edges which are j steps away from the node i. Since we are considering degree of all nodes and weights of all edges, usually the total power of all nodes will be different except for vertices which are same with respect to an isomorphism. When β value is very large, this method tends to usual degree centrality. We rank the nodes in the descending order of total power $T(i)$. If we put $\alpha = 0$, this method can be applied to unweighted networks. In Ref. [18], the authors have verified the reliability of this method using rank correlation method.

PageRank [19] is a method for rating web pages effectively by measuring the human interest and attention devoted to them. In the network of web pages, a webpage is a node and a hyperlink is a directed link in the network. It is the ranking method used by Google. In PageRank, a hyperlink is understood as an endorsement relationship. In this method, number and quality of links to a page are used to determine a rough estimate of how influential the web page is. This method was first introduced in 1976 by Gabriel Pinski and Francis Narin. The PageRank is defined as a function

$$PR(T_A) = (1-d) + d\left(\frac{PR(T_1)}{C(T_1)} + \ldots + \frac{PR(T_n)}{C(T_n)} \right) \quad (19)$$

where $PR(T_A)$ is the page rank of T_A and $PR(T_i)$ is the page rank of T_i which is linked to T_A. $C(T_i)$ is the number of outbound links on page T_i, and d is a parameter between 0 and 1.

3. Conclusion

A large variety of node centrality measures are being used by network scientists to identify potential spreaders of information in a network. Degree centrality, betweenness centrality, closeness centrality, k-core decomposition, mixed degree decomposition, improved k-shell decomposition, weighted k-shell decomposition, page ranking and m-ranking method, etc., are important methods, which are discussed in this paper. These methods help us to rank the members in a network according to their importance. A method, which is suitable in some context, may not fit in another situation. Each method has its own strengths and weakness. The weakness of the methods motivates social scientists to redefine or modify the existing methods, which in turn leads to the development of new methods to rank the nodes in a better way. All the methods, except m-ranking method, have been criticized about its inefficiency in uniquely ranking the members. In addition, some methods may give higher rank to less important nodes, while most important nodes being underestimated. It has been claimed that m-ranking method is far better than all other methods, which are described above.

Acknowledgements

This research was completed with the financial assistance of University Grants Commission, India. First author is grateful to University Grants Commission for sanctioning a major research project titled, "Modeling of social ties; a modified approach", No. F. No. 40-243/2011 (SR).

Author details

Reji Kumar Karunakaran[1]*, Shibu Manuel[2] and Edamana Narayanan Satheesh[3]

*Address all correspondence to: rkkmaths@yahoo.co.in

1 Department of Mathematics, N. S. S College, Pandalam, Kerala, India

2 Department of Mathematics, St. Dominic's College, Kanjirapally, Kerala, India

3 Department of Mathematics, N. S. S. Hindu College, Changanacherry, Kerala, India

References

[1] Barrat A, Barhlemy M, Vespignani A. Dynamical Process on Complex Networks. Cambridge University Press, Cambridge 2008

[2] Newman MEJ. Spread of epidemic disease on networks. Physical Review E. 2002;**66**(1): 016128

[3] Cohen FB. A short Course on Computer Virus. New York: John Wiley and Sons; 1994

[4] Kephart JO, Sorkin GB, Chess DM, White SR. Fighting computer viruses. Scientific American. 1997;**277**(5):56–61

[5] Arnaboldi V, Conti M, Pezzoni F. Ego network structure in online social networks and its impact on information diffusion. Computer Communications. 2015. DOI: 10.1016/j.comcom.2015.09.028

[6] Nekovee M, Moreno Y, Bianconi G, Marsili M. Theory of rumour spreading in complex social networks. Physica A. 2007;**374**:457–470

[7] Rosnow RL, Foster EK. Rumor and gossip research. 2005; American Psychological Association (www.apa.org/science/about/psa/2005/04/gossip.aspx)

[8] Zeng A, Zhang C. Ranking spreaders by decomposing complex networks. Physics Letters A. 2013;**377**:1031–1035

[9] Liu J, Ren Z. Ranking the spreading influence in complex networks. Physica A. 2013;**392**: 4154–4159

[10] Bae J, Kim S. Identifying and ranking influential spreaders in complex networks by neighborhood coreness. Physica A. 2014;**395**:549–559

[11] Newman MEJ. Networks: An Introduction. Oxford, UK: Oxford University Press; 2010

[12] Opsahl T, Agneessens F. Node centrality in weighted networks: Generalizing degree and shortest paths. Social Networks. 2010;**32**:245–251

[13] Freeman LC. Centrality in social networks conceptual clarification. Social Networks. 1979;**1**:215–239

[14] Garas A, Schweitzer F, Havlin S. A k-shell decomposition for weighted networks. New Journal of Physics. 2012;**14**:083030

[15] Wei B, Liu J, Deng Y. Weighted k-shell decomposition for complex networks based on potential edge weights. Physica A. 2015;**420**:277–283

[16] Reji Kumar K, Manuel S. Spreading information in complex networks: A modified method. In: Proceedings of International Conference on Emerging Technological Trends 2016. ICETT 2016 Organized by Noorul Islam University, Thakkalai, Tamil Nadu, India. To appear in IEEE Digital Explore Library

[17] Wang Z, Zhao Y. Fast ranking influential nodes in complex networks using a k-shell iteration factor. Physica A. 2016;**461**:171–181

[18] Reji Kumar K, Manuel S. The m-ranking of nodes in complex networks. In: Proceedings of COMSNETS 2017. 9th International Conference on Communication Systems & Networks, Bangalore, India, during January 4-8, 2017. To appear in IEEE Digital Explore Library

[19] Franceschet M. PageRank standing on the shoulders of giants. ACM. 2011;**54**:92–101

Chapter 2

An Example Usage of Graph Theory in Other Scientific Fields: On Graph Labeling, Possibilities and Role of Mind/Consciousness

Auparajita Krishnaa

Additional information is available at the end of the chapter

http://dx.doi.org/10.5772/intechopen.68690

Abstract

This paper provides insights into some aspects of the possibilities and role of mind, consciousness, and their relation to *mathematical logic* with the application of problem solving in the fields of psychology and graph theory. This work aims to dispel certain long-held notions of a severe psychological disorder and a well-known graph labeling conjecture. The applications of graph labelings of various types for various kinds of graphs are being discussed. Certain results in graph labelings using computer software are presented with a direction to discover more applications.

Keywords: mathematical logic, graph labeling, magic, antimagic, inner magic, inner antimagic, graceful, harmonious, felicitous, sequential, NP-Complete

1. Introduction

In this paper, we begin with looking at the problems and problem solving elements of *mathematical logic* of the field of Discrete Mathematics (of which graph theory is a subject) and its application in the realms of mind and consciousness. Section 2 focuses on dispelling certain notions about a severe psychological disorder namely *paranoid schizophrenia*. It is a much debated and often misunderstood illness affecting many people. Many remain untreated and undiagnosed not even able to adequately articulate their thoughts and feelings causing them a severe debility and an untold suffering.

Generally, the mind and consciousness aspects have been nearly left out from Mathematics for various reasons but mostly studied in other fields. Perhaps, the first instance (as per the available literature) in solving a major problem in Mathematics providing an answer using *mathematical arguments* factoring in the role of mind/consciousness is given in Ref. [1] resolving the *graceful tree conjecture*. In Section 3, we look at some preset notions in assuming this conjecture to be true and some new insights into the work in Ref. [1]. This conjecture was widely believed to be true for about five decades that all trees are graceful. This conjecture was believed to be true by many researchers in the field of Graph Theory particularly in graph labelings even without proof. A *graceful labeling* in a *graph* with p vertices(points) and q edges(lines) is assigning the numbers (labels) to the vertices $0,1,2,\ldots,(q-1)$ such that the induced edge labels found by taking the absolute values (the positive value) are from $1,2,\ldots,q$ (**Figure 1**). This is an interesting area of Graph Theory as well as Psychology to be studied that why this conjecture was believed to be true even without proof for such a long time and that too in the field of Mathematics which is often considered by many to be so cut and dry in terms of proofs and truth of ideas.

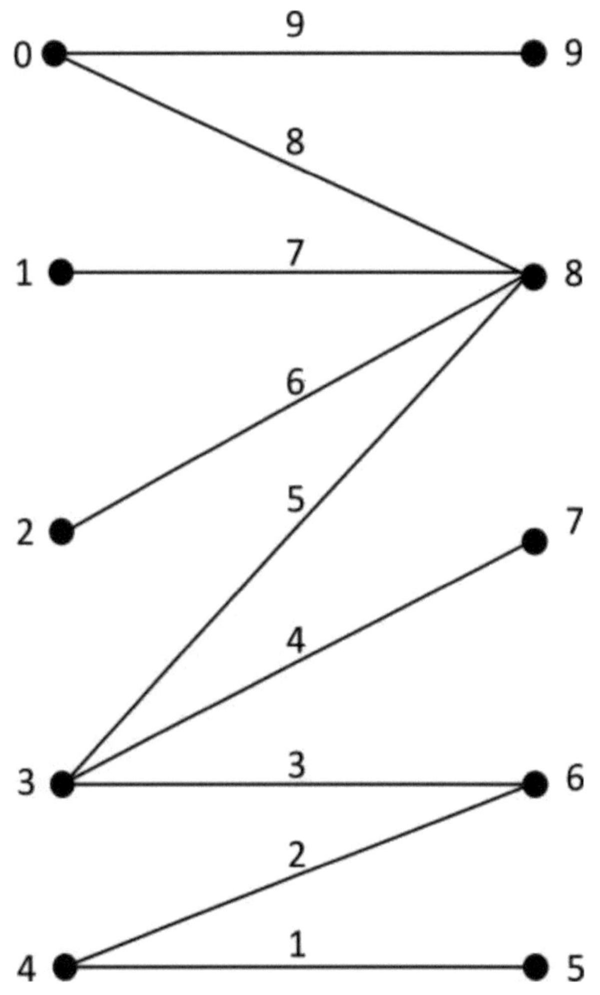

Figure 1. Graceful bipartite tree where edges do not cross.

In Section 4, certain new aspects of applications of *antimagic, inner magic,* and *inner antimagic* labelings and other kinds of labelings given in Refs. [2, 3] are presented. The *inner magic* and *inner antimagic* are new kinds of labelings given in Ref. [4]. In Section 5, some results using computer software in the field of graph labelings are discussed. In Section 6, a discussion of the *lexicographic order* in graph labelings is presented with a direction to future study. At last, conclusion is given in Section 7.

2. Resolving aspects of paranoid schizophrenia

During the course of one's life, one comes across various psychological problems, sometimes of a severe nature of psychological disorder like *schizophrenia. Schizophrenia* has perplexed researchers, thinkers, doctors, philosophers for centuries, though the systematic study of *schizophrenia* is approximately 100 years old and has been called "arguably the worst disease affecting mankind, even AIDS not excepted (Nature 1988)" in Ref. [5]. In this section, we would focus on understanding *schizophrenia* particularly *paranoid schizophrenia.* The clinical symptoms are characterized by the following major symptoms as in [5–7].

1. Delusions of persecution like the person believing others conspiring against him/her or grandiosity.
2. Delusion of exalted birth like the belief that one is born with a messianic role.
3. Delusions of reference like believing that television, newspapers are referring to the patient in particular.
4. Delusions of jealousy/infidelity: hearing voices of threatening nature which give commands to patient.
5. In Ref. [7], "People with schizophrenia often have delusions, firmly held, unshakable beliefs with no basis in reality."
6. Hallucinations ("experience of perceiving things that do not actually exist" in [7]) of non-verbal auditory nature like laughing, humming, etc.
7. Hallucinations through other somatic sensations like smell, sight, taste, etc.
8. Withdrawal or hesitation from interacting with other people and feeling isolated.
9. In Ref. [5], "The delusions are usually well-systematized (i.e., thematically well connected with each other)."
10. In Ref. [5], under the first rank symptoms (SFRs) of *schizophrenia.*
    (a) Voices heard arguing: two or more hallucinatory voices discussing the subject with third person.
    (b) 'Made' volition of acts: In made effect, impulses and volitions, the person experiences feelings, impulses, or acts which are imposed by some external force. In "made"

volition, for example, one's own acts are experienced as being under control of some external forces.

(c) Somatic passivity: Bodily sensations, especially sensory symptoms are experienced as imposed on body by some external force.

(d) Delusional perception: Normal perception has a private and illogical meaning.

11. Ref. [6] gives three types of *thought alienation* under Schneiderian first rank symptoms of *schizophrenia* as follows:

(a) Patient believes that his/her thoughts are under control of some external agency and others participate in this thinking.

(b) External thoughts are being put in his/her mind by some external force.

(c) *Thought broadcasting*: patient feels that his/her thoughts are read by others and broadcast.

12. In Ref. [7], "schizophrenia is a form of regression to earlier experiences and stages of life."

13. Patient can become homeless, suicidal, unemployed and behavioral problems could be bizarre.

Paranoid schizophrenia is the most common type of *schizophrenia,* and the patient may not appear psychiatrically ill unless the symptoms of paranoia come to light.

Every person is conditioned by one's culture, educational training, ethics, environment, life events, etc. Often this illness is treated by the psychiatrist based on his/her medical training, experience, and often colored by the above mentioned conditioning with perhaps their own perception that the present knowledge about this illness is all there or just plain puzzled compounded by the fact that every patient is similar and different as well in many ways. There is much more to these hallucinations, delusions, and symptoms from the mathematical and scientific viewpoints and also there are certain *bases and reasons* of these symptoms which are often not understood adequately.

Often Nature's mysteries present themselves in form of problems to be resolved with deeper hidden truths contained in them. We discuss the above mentioned aspects of hallucinations and delusions using principles of *mathematical logic* followed by explanation and insights into these symptoms and this illness.

Theorem 1: Delusions and hallucinations exist from an objective point of view in *paranoid schizophrenia.* This is proven as follows:

Prove the validity of the argument:

If Neela is a patient of *paranoid schizophrenia,* then she has delusions and hallucinations. If delusions and hallucinations are experienced by Neela, then these are either considered "unreal" (imaginary) by Neela's doctors or delusions, hallucinations are considered "real"

http://dx.doi.org/10.5772/intechopen.68690

(existent) by Neela. Either delusions or hallucinations are not considered "unreal" by her doctors (i.e., cause is found or insight into certain traumatic event(s) causing them and issue resolved) or no debility, behavioral problems caused by hallucinations and delusions remain (i.e., way of solution found to understand delusions and hallucinations of the *particular patient*) or delusions and hallucinations exist from an objective point of view in *paranoid schizophrenia*. Therefore, delusions and hallucinations exist from an objective point of view in *paranoid schizophrenia*.

Proof: First, we convert the given argument in symbolic form using the *propositional variables a,b,c,d,e,f* and *logical connectives such as and* (^), *or* (v), *not* (¬), *implication if-then* (→):

a: Neela is a patient of *paranoid schizophrenia*.

b: She has delusions and hallucinations.

c: Delusions and hallucinations are considered "unreal" by Neela's doctors.

d: Delusions and hallucinations are considered "real" by Neela.

e: Debility and behavioral problems are caused by delusions and hallucinations.

f: Delusions and hallucinations exist from an objective point of view in *paranoid schizophrenia*.

The given argument now becomes in symbolic form as follows:

a → b (*If* Neela is a patient of *paranoid schizophrenia then* she has hallucinations, delusions)
b → (c ^ d) [*If* she has delusions, hallucinations *then* {(these are considered "unreal" by her doctors) *and* (these are considered "real" by Neela)}]
¬ c v (¬ e v f) [delusions, hallucinations are *not* considered "unreal" by her doctors *or* {(*no* debility, behavioral problems caused by delusions, hallucinations remain) *or* (delusions, hallucinations exist from an objective point of view in *paranoid schizophrenia*)}]
a ^ e (Neela is a patient of *paranoid schizophrenia and* debility, behavioral problems are caused by delusions, hallucinations)

Therefore, f (Delusions and hallucinations exist from an objective point of view in *paranoid schizophrenia*).

Steps	Reasons
1. a → b	Premise
2. b → (c ^ d)	Premise
3. a → (c ^ d)	Steps 1 and 2 and the law of syllogism
4. a ^ e	Premise
5. a	Step 4 and rule of conjunctive simplification

Steps	Reasons
6. c ^ d	Steps 5 and 3 and rule of detachment
7. c	Step 6 and rule of conjunctive simplification
8. ¬ c v (¬ e v f)	Premise
9. ¬ (c ^ e) v f	Step 8, the associative law of 'v' and De'Morgan's law
10. e	Step 4 and rule of conjunctive simplification
11. c ^ e	Steps 7 and 10 and rule of conjunction
12. Therefore, f	Steps 9 and 11 and rule of disjunctive syllogism

Thus, we see that the delusions and hallucinations exist from an objective point of view in *paranoid schizophrenia*. The rules of *mathematical logic* are in Ref.[8].

Often the hallucinations and delusions not being believed to be "real" by the doctors and the patient unable to verbalize the dreadful feelings contained in the delusions and hallucinations lead to the patient feeling helpless and hopeless. Everything which exists in nature has an element of truth in it and is very much "real" which can appear as something quite negative until not understood. Nature presents problems to uncover the hidden truth underlying the *causes* in a problem. We would look at the causes of development of hallucinations and delusions which are the major symptoms of *schizophrenia*.

There is often the basis of traumatic abusive events from distant past behind hearing the voices and delusions. Some abusive traumatic events by an older person in position of power and trust can get buried in the memory of the patient and get internalized as "just punishment" or even "love" thus making it seems like "normal." This "normal" can create later on conflict in the personality between "right" and "wrong" accompanied by tremendous guilt, fear, anger, self-doubt, and paranoia. Oftentimes, the acts of abuse including rapes are carried out in childhood or during adulthood in a vulnerable situation of the victim by cajoling *suggestions* by the abuser with *cajoling* seeming like "love" to the victim or in a threatening manner causing terror in the victim. In either case such events create a lot of guilt, fear, and paranoia in the psyche of the victim. The power of a loving suggestion is not lost on anyone but these very suggestions to carry out a crime can be internalized very deeply by the victim and could be heard as **"voices"** later in life and being under **"control" or "persecuted" by some external agent**. Such cunning suggestions can also have the effect of *hypnosis* on the victim with the victim not realizing that he/she has fallen prey to a heinous crime. The whole incident(s) can lie buried in the consciousness of the victim for a long time like a volcano which can erupt anytime triggered by some sensory input from the outer environment, especially sound. For example, in distant past, some violent incidents were followed by the banging of utensils and food served to the patient in those utensils. In the present moment, the patient's dormant memory can get horribly triggered by sound of some utensils or similar sound causing ***auditory hallucinations***. Moreover, the victim can pick up the opinions and feelings of the abuser at the time of crime. For instance, picking up of *mocking attitude* of the abuser can result in hearing voices or hallucinations of "**nonverbal auditory nature**

http://dx.doi.org/10.5772/intechopen.68690

like *laughing*" at a later date by the patient. The patient can feel that the abuser (the personality) has **"entered"** his/her mind-body system. This can cause confusion over time and an apparent incoherent jumbled up speech of the patient and having hallucinations. The patient can feel **"possessed"** by the person who committed the crime on him/her or in case of not remembering accurately, the patient can feel the same by projecting his/her experience to another person not responsible for the crime. In this way, the patient can feel that **"external thoughts are being put in his/her mind by some *external force*."** Being termed the hallucinations as "unreal" or *"which actually do not exist"* by the doctor does not help the patient except increasing the terror felt by the patient and losing heart even further. Usually crimes of this nature have certain intimacy creating emotional dependence between the abuser and the victim. When such incidents are not discussed immediately or soon after the incidents, the thoughts and feelings tend to fester inside the mind-body of the patient and *not forgotten in the real sense* but deeply suppressed and internalized as part of the personality of the patient. The added guilt also comes from the victim being blamed for crimes of this nature for various socio-cultural reasons. The stigma and taboo associated with such crimes and mental disorders complicate the situation even further.

The abusive events and their memories keep on getting buried deeper and deeper in the consciousness of the victim with time, building up of ***avoidance issues*** and sucking up the psychic energy of the mind-body system causing debility. The victim tends to develop ***thought patterns*** of escaping the trauma and tends to get attracted to similar abusers inviting trouble for him/her and over long period of time, the thought patterns, feelings, events getting all mixed up developing hallucinations and delusions. ***If*** the hallucinations and delusions did not actually exist and were **"imaginary," then** there would be ***no debility***. The patient has an **"unshakable belief"** that the delusions and hallucinations are true through subconscious memory and knowledge but cannot articulate them due to the memories being overwhelming. Thus, the building up of trauma and thought patterns can go on till death and beyond. During intense emotional pain, the patient may realize that the psyche has very much a life of its own and certain past life event may **resonate** with similar event in the present lifetime. As in Ref. [7] "**schizophrenia is a form of regression to earlier experiences and stages of life."** Whether one believes in past lives or not depends upon how convincingly one can recall certain events. Going back in one's life may be understood in this way that if one is an adult then he/she has been a child also and some traumatic event(s) must have caused the delusions and hallucinations. A particular patient may have the ability to regress to childhood trauma or could regress further to pre-birth times as well. The occurrence of past lives has got a lot to do with the patient's ability to recall correctly and can be understood from the **broader viewpoint of nature's perspective**. Nature creates, destroys, and maintains the creation through **dualities** like right-wrong, pleasure-pain, day-night, birth-death, etc. The body has a shelf-life of a few decades and, in general, has limitations of movement but the psyche is not limited like the body—its life is indefinite. Psyche is more powerful than the body and more subtle. It has more potential for evolution in terms of **self-knowledge**, though more complex and difficult than the physical. When the body is done with the death, the invisible psyche remains because mind/consciousness is a form of ***invisible but palpable energy***. The psyche has a subconscious longing to resolve the pending issues of the trauma and is often full of regret. This psyche assumes next body with mixed emotions with the already developed thought patterns and unresolved trauma. In the present lifetime of the patient, some memory may be triggered off by some stimuli or coming across the abuser in a different body and the patient may feel that the **"sensory symptoms are imposed on body by some external force."** When looked at from a broader time-span, it can be understood that the psyche is a form of energy, only with various kinds of thoughts, feelings, and with some **unfinished business**. One can have restlessness due to some *unfinished business* for a few days in the same body then why not for a longer time-span? Thought acts as ***chronological time*** in the mind. It can be understood in this way that taking urgent action immediately which someone was thinking of taking after 2 years takes away those 2 years, and one feels unburdened thus bringing the "future" into the "present." Some past memories can get triggered *now* which bring the "past" into the "present." Therefore, the thoughts and feelings dictate to a great extent the interpretations of *past, present,* and *future.* So, the ***psychological time*** is different from the *chronological time.* Some gifted patients can recall the past lives in a "timeless" way i.e., by looking at his/her life events as events of one stretched out lifespan with events well connected and with insights into the thought patterns (as in no. 9 of symptoms: "***the delusions are usually well-systematized i.e., thematically well connected with each other***"). This is entirely possible if the patient, who has authentic paranormal abilities, is a conscientious, courageous warrior, and a lover of truth.

The intelligence of the mind-body complex records everything like a movie or even better. The dialogues between the abuser and the victim may appear as "voices" later on in life or if the person is sensitive may be constantly reminded of it. Based on some similar stimuli in the outside environment, this/these traumatic event(s) stored in the mind-body can get triggered off. Usually, the victim is blamed directly or indirectly thus causing a severe guilt and fear which when accumulated over time can result in **"delusion of reference, persecution"** or simply projecting his/her guilt and fear to everyone outside thus resulting in **"thought broadcasting."** Guilt and low self-esteem can only intensify the feeling of **"others will read my thoughts and harm me."** An abusive incident can breed the feelings of betrayal, infidelity, jealousy, insecurity, feelings of no sense of belonging to anyone or anywhere and broken-hearted leading to a solidified mental state of feeling orphaned, and **"homeless."** This could translate to the patient actually running away from home, hospital, or refusing treatment. The patient may also "learn" infidelity attracting similar abusive people and "expecting" to be betrayed which may appear as "natural" in relationships and could cause major relationship troubles. Memories recorded in the consciousness of such abusive incidents are experienced as ***hallucinations of a threatening nature giving commands to the patients***; these are very real to the patient but fear of not being believed and unable to talk about them only makes things worse because of guilt associated with infidelity and anger with jealousy.

A gifted individual can be subjected to actual events of ostracizing, persecution, and condemned for his/her free ideas not fitting in with the majority of people, so the *delusions of persecution* can be quite pronounced in this case with the patient feeling like a criminal burdened with fear, guilt, and a seething insane rage underneath. If he/she has certain fame due to excellence in work, then the **delusion of reference** can bother him/her like anything.

http://dx.doi.org/10.5772/intechopen.68690

A patient can pick up certain disturbing *opinions* common to large number of people (*collective consciousness*) either through hearing, reading about them, watching a movie, etc. which caused much pain to the patient and he/she can feel **"attacked"** by several people triggering off hallucinations and delusions. In terms of picking up others' thoughts and feelings the illness of *schizophrenia* can get as **"infectious"** or more than a communicable illness of the body.

In the case of an average patient or a gifted patient, abuse at a vulnerable situation can happen, and the insight into the truth of such events can involve strong emotions like betrayal, anger, and helplessness. The prospect of a daunting war to be fought within and with the abusers who could be far away coupled with the paralyzing stored up guilt, terror, anger, sorrow, debility, etc. may depress the patient like anything. This is the trial by fire and the test of inner strength of character which the patient may face. Moreover, an intelligent action of the medicine could try to release the suppressed traumatic memories (this may be understood as being akin to an operation on the body to take out a tumor, only in the psychological realm, it is much more complex and difficult). This can seem like worsening the situation thus the patient losing faith in the medicine, stopping it, and relapses taking place. An ardent wistful desire to avenge and revenge could be behind the ***delusion of grandiosity*** **and feelings of one's messianic role in life**. Even when not triggered, the person may hear voices subconsciously of those stored events in the memory which are so old that the memory appears to be lost and cannot be recalled. If it could be recalled or some insight into the thoughts and feelings happen, then it could lead to recognizing and changing the *thought patterns* by the patient with suitable help. But the essential initiative must come from the patient and could lead to decrease in delusions and hallucinations and hence improvement in behavioral problems. As in Ref. [5], **"The delusions are usually well-systematized, i.e., thematically well connected with each other"** also points to the fact that delusions **stem from some events** and the need to start by believing the patient could be the first step instead of right away starting with the notion that voices or delusions are imaginary or "unreal."

Even in the case of delusions and hallucinations being "unreal," we have to see that **in nature or creation everything appears as *dualities or pairs of opposites*, and there is *truth* in both the extremes. Change or evolution which is continuous is brought about by these pairs of opposites by nature in creation**. For instance, like pleasure-pain, day-night, birth-death, true-false, for-against, etc., the "real"-"unreal" in this case is also another pair of opposites with truth in both, depending on whose viewpoint one is looking from, the puzzled doctor, or the troubled patient. In shaping of the consciousness of a person, the elements affecting the psychology like personal background, the kind of education one has had, culture, ethics, events in one's life, etc. all play a role in shaping and conditioning the mind, thus the doctor is also conditioned by these factors as well as the patient. Moreover, these factors are *unique* for each person hence for each patient as well; therefore, the patient in order to get well has to take the initiative himself/herself to understand this conditioning with suitable help.

One more thing to be noted is the overemphasis in human society on the logical part and not emphasizing enough in the matters of heart. In climbing the social ladder, the calculative logical part plays a greater role disregarding the heart/conscience which speaks our truth to us, and this voice is ignored often creating transgressions against the conscience leading to dishonesty, obscuring of the truth, and further complicating the disorders of the psyche. Another outcome of overemphasis on logic in terms of social conditioning is that when due to some input from the senses, some trauma is caused or triggered; the person is unable to feel and analyze the overwhelming emotions as emotions are far more powerful than thoughts and could lead to bizarre behavior. This conditioning is equivalent to total disregard of the heart leading to ignoring the heart's voice further leading to an *unbalanced personality*.

There is often deep depression, exhaustion, anxiety, panic, post traumatic stress leading to psychosis, obsession, and one, more, or all of these are experienced by the patient at some time or other in *schizophrenia* which makes it difficult to diagnose and understand this illness.

The problems in the psychological realm get complicated by the fact that the world is flooded with the advice of many experts on "positive thinking" which is an antithesis to the approach of *problem solving*. This *positive thinking* often translates into an avoidance, even denial, looking the other way and away from the problem. The **apparently negative approach to problem solving requires** hard work, paying attention to the problem and taking adequate actions in response to various psychological states, thoughts and feelings and is anything but escapist. Not working on *problem solving* only aggravates the problem.

About the correlating of thoughts and ideas**, Ref.** [5] says "Autistic thinking is one of the most classical features of schizophrenia. Here, thinking is governed by private and illogical rules. The patient may consider two things identical because they have identical predicates or properties (**von Domarus Law**)." For example, Jesus Christ was persecuted, I am persecuted; So I am Jesus Christ.

It is a matter of interest to see the similarity between the above law and the **Law of Detachment in *mathematical logic*** of Discrete Mathematics. Let us consider the same argument in light of the law of detachment as follows:

I feel persecuted. If I feel persecuted then I am Jesus Christ. Therefore, I am Jesus Christ.

In symbolic notation, the argument using the *propositional variables* is as follows:

p:I feel persecuted.

q:I am Jesus Christ.

Let us write the argument as per the Law of Detachment in symbolic notation as follows:

p (I feel persecuted)

$p \rightarrow q$ (*If* I feel persecuted *then* I am Jesus Christ)

Therefore, q (Therefore, I am Jesus Christ).

The Law of Detachment is a *valid rule of inference* in *mathematical logic*. This similarity points to the ***common origin*** of *logic/arguments* in Psychology and Mathematics. Thus, we observe that what appears as **"illogical"** to the doctor is a perfect valid argument mathematically and psychologically for the patient further establishing our findings in this section depending upon whose viewpoint one is looking from. In fact, the whole of the discussion in this section is of *mathematical logic* (*reasoning*) applicable in psychological realm as well. The patient is obviously not Jesus Christ from the doctor's point of view but the patient unable to feel or understand the trauma **identifies** with the trauma he/she knows as what Jesus Christ faced and expresses in this manner.

Schizophrenia is an illness of facing the hard truths in one's life or lifetimes as the case may be depending on the patient's abilities, initiative, and sense of responsibility toward oneself preferably with suitable help.

3. Role of mind and consciousness in resolving the graceful tree conjecture

Sometimes holding on strongly to the binary logic of true/false in Graph Theory (a branch of Discrete Mathematics) fails to produce results and can even make mathematics look too rigid. This has been seen in the case of the well-known about five decades old *graceful tree conjecture* (all trees are graceful) being assumed to be true by a large number of researchers even without proof.

Tree is a kind of *graph*. A *graph labeling* is assigning numbers (labels) to vertices and edges such that the induced labels form a certain pattern. A *graceful labeling* in a graph with p vertices and q edges is assigning the numbers to the vertices $0,1,2,\ldots,(q-1)$ such that the induced edge labels found by taking the absolute value (the positive value) are from $1,2,\ldots,q$ (see **Figure 1**). Ref. [9] was an attempt to solve the graceful tree conjecture. Ref. [9] was a *direct attempt* to solve the graceful tree conjecture as pointed out in Ref. [10].

Assuming this conjecture to be true led to no doubt a lot of work getting done and published but generally of specialized variety and not able to pinpoint why and how this conjecture was even formed in the first place. Zeroing in on answering this conjecture in Ref. [1] after attempting to solve it in Ref. [9] took years of delving into the nature of the mathematical problem as well as the nature of mind to discover that it seemed that the definition of tree being *connected and acyclic* was taken to be the controlling factor in determining that *all trees* would be *graceful*.

A *bipartite graph* is one whose set of vertices can be split into two subsets X and Y such that each edge of the graph joins a vertex in X and a vertex in Y. The following *algorithm* G in Ref. [9] gives *graceful labeling* for ***bipartite trees* in which the edges do not cross** (**Figure 1**).

Algorithm G

---

1. Draw the tree as a bipartite graph in two partite sets denoted as Left(L) and Right(R). Let the number of vertices in L be x.
2. Number the vertices in L starting from top going to bottom consecutively as $0,1,\ldots,(x\text{-}1)$.
3. Number the vertices in R starting from bottom going to top consecutively as $x,(x+1),(x+2),\ldots,q$ (q is the no. of edges). Note that these numbers are the vertex labels.

4. Compute the edge labels by taking the absolute value of the difference of the incident vertex labels.

5. The resulting labeling is graceful.

The *trees* which cannot be drawn as *bipartite trees* where the edges do not cross may not give *graceful labeling* with this algorithm. Other algorithms in [9] for labelings of such bipartite trees namely ***harmonious*** (*vertex labels are added mod q; one vertex label allowed to be repeated in trees*: **Figure 2**), ***sequential*** (*vertex labels are added with ordinary addition*: **Figures 2** and **3**), **felicitous** (*vertex labels are added mod q*: **Figure 3**), *and* **antimagic labelings** (*sum of all the incident edge labels at each vertex is distinct*: **Figures 4** and **5**) are also based on this construction of *trees*.

In a result like this, the mind and its set perceptions play a role because the mind being conditioned by a long history of the traditional branches of mathematics where the mathematical structures are rigid, well defined and are controlled or predicted by the definitions, tends to **assume or project the same** in Graph Theory as well. In geometry also there are diagrams but these diagrams are defined by their formulas. For instance, ellipse, hyperbola, etc. are defined by their formulas or how many geometrical units will comprise a given

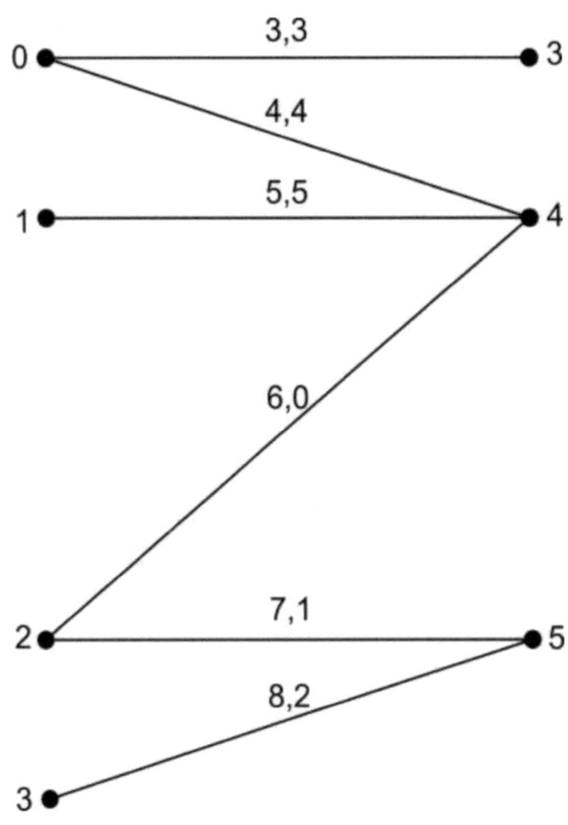

Figure 2. Sequential and harmonious bipartite tree where edges do not cross. Sequential and harmonious edge labels written on left and right, respectively, separated by a comma.

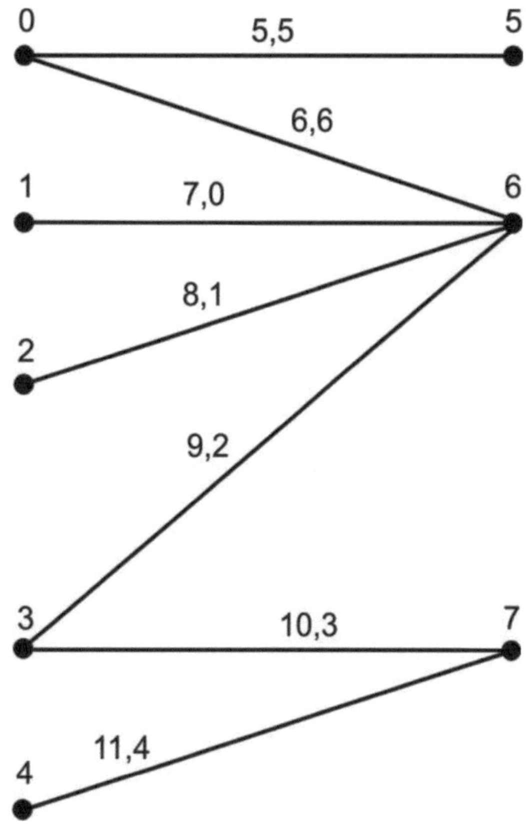

Figure 3. Sequential and felicitous bipartite tree where edges do not cross: Sequential and felicitous edge labels written on left and right, respectively, separated by a comma.

diagram is specified as in a triangle (three line segments and angles), square, etc. But *trees* are just defined as *connected and acyclic* and can be drawn in any way with ***no restrictions and nothing which can be measured or quantified resulting in infinite shapes*** rendering them suitable only for specialized results of particular kinds of *trees* and not generalized results. So far, the research in Mathematics has discounted how the mind influences the perceptions of looking at mathematical problems, formulating them, and going about solving them. How mind and its perceptions influence viewing a mathematical problem, its formulation and solution could be an area of research which can be quite subjective but could yield interesting results when done more consciously. Ref. [1] which answers the *graceful tree conjecture* and Ref. [11] which finds that the definition of tree alone can not be the controlling factor in similar conjectures, both factor in the psychological conditioning in looking at a mathematical problem.

From the rules of *mathematical logic* some of above mentioned ideas can be proved as follows:

Theorem 2: Definition of *tree* alone is not the controlling factor in assuming the *graceful tree conjecture* to be true.

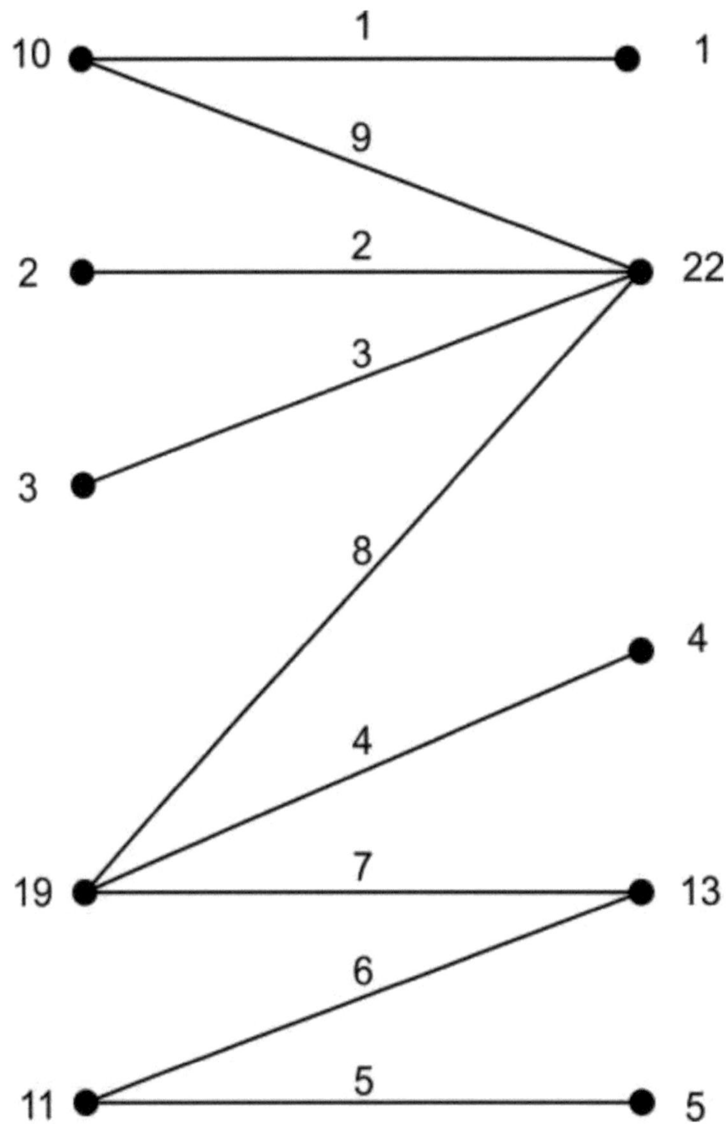

Figure 4. Antimagic bipartite tree where edges do not cross.

Proof: Let us write the argument using *propositional variables* and *logical connectives* as follows:

p: definition of *tree* alone is the controlling factor in assuming the *graceful tree conjecture* to be true.

q: graceful tree conjecture is assumed to be true By the Law of Modus tollens the argument is as follows:

$p \rightarrow q$ (*if* definition of *tree* alone is the controlling factor *then the graceful tree conjecture* is assumed to be true).

$\neg$ q (*graceful tree conjecture is not* assumed to be true).

http://dx.doi.org/10.5772/intechopen.68690

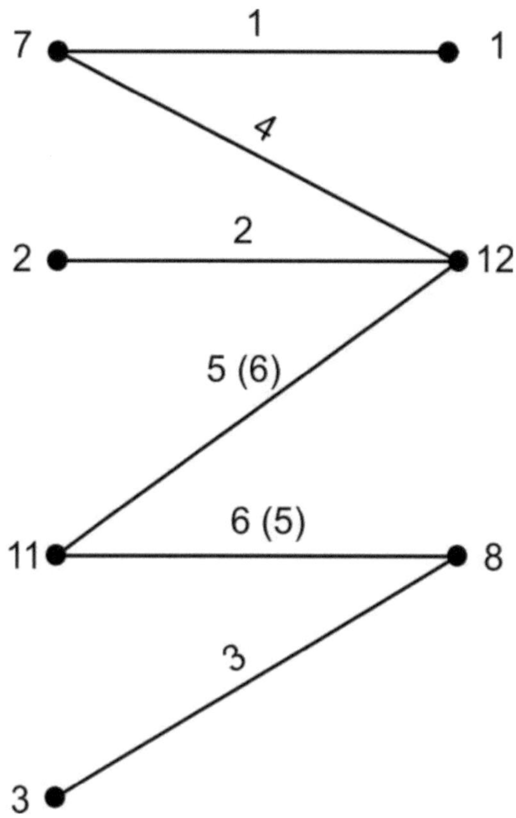

Figure 5. Antimagic bipartite tree where edges do not cross: the edge labels in parentheses show the interchanged edge labels as in step 4 of algorithm AM.

Therefore $\neg$ p (definition of *tree* alone is *not* the controlling factor in assuming the *graceful tree conjecture* to be true).

Thus, it should be noted that pictorial representation of graphs gives rise to infinite number of shapes which makes impossible to classify them all and makes the definitions in Graph Theory particularly in *trees*, distinct from definitions in older, traditional branches of Mathematics, and may not be the basis of forming generalized conjectures like the *graceful tree conjecture.*

4. Graph labelings and applications

As from the *antimagic, inner magic,* and *inner antimagic* graphs, applications have emerged and been presented in Ref. [2], and some new and modified interpretations are being given in this work. *Inner magic* and *inner antimagic* are new kinds of labelings given in Ref. [4]. In these labelings, the p vertices, q edges, and the f internal faces of a planar graph are labeled such that labels of the faces form an *arithmetic progression* with *common difference* d. If $d = 0$, then the graph is said to have an *inner magic* labeling, and if $d \neq 0$, then the graph is said to have *inner antimagic* labeling. **Figure 6** shows the *inner magic* and *inner antimagic* wheel.

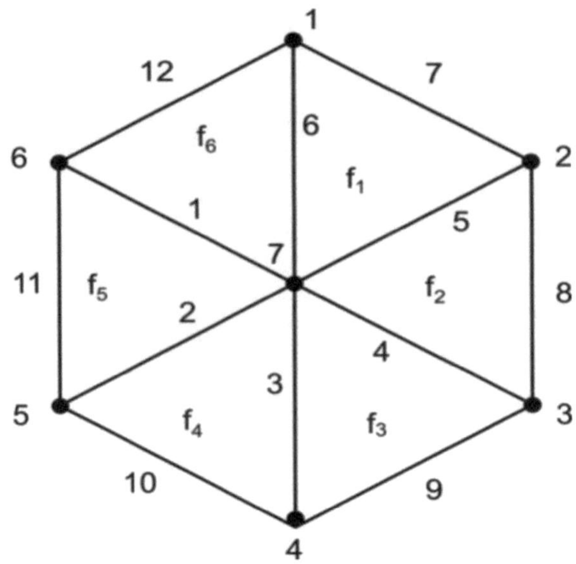

Figure 6. Inner magic and inner antimagic wheel. Inner magic internal face labels: $f_1 = 6, f_2 = 5, f_3 = 4, f_4 = 3, f_5 = 2, f_6 = 1$. Inner magic weight number = 34. Inner antimagic internal face labels: $f_1 = 1, f_2 = 2, f_3 = 3, f_4 = 4, f_5 = 5, f_6 = 6$. Inner antimagic internal face weights: 29, 31, 33, 35, 37, 39.

An *antimagic* labeling of a graph with *p* vertices and *q* edges is one in which the *q* edges are labeled with numbers 1,2,…,*q* such that the sum of the incident edge labels on each vertex is distinct. In the *antimagic* graphs shown in **Figures 7–11**, the vertices could represent offices in a building and the *antimagic* labeled edges could represent codes/passwords to reach those offices. The *antimagic* label of the particular vertex representing office could be a password or code given

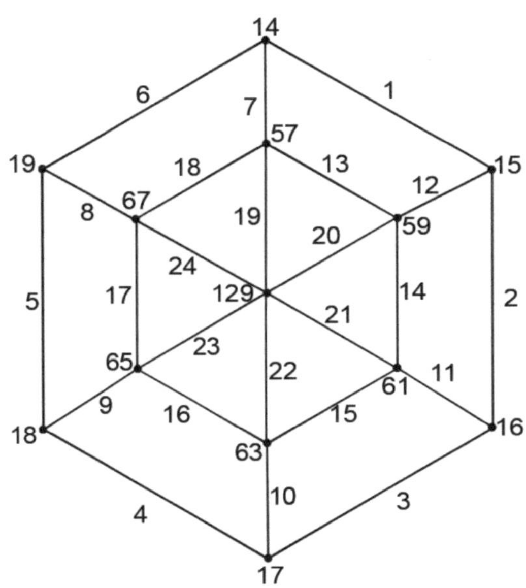

Figure 7. Antimagic double wheel D_6.

http://dx.doi.org/10.5772/intechopen.68690

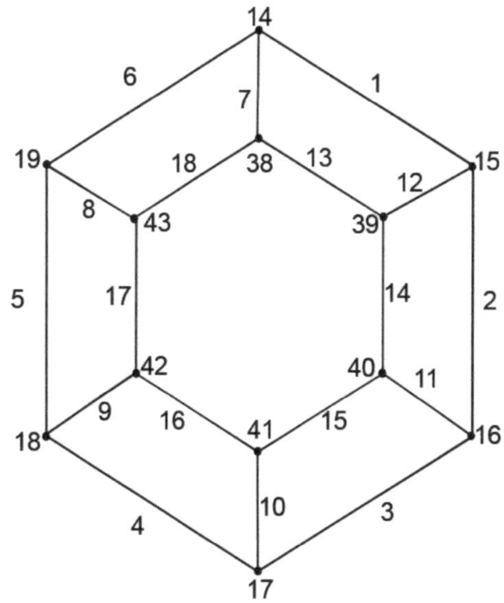

Figure 8. Antimagic centerless wheel-2 $CW2_6$.

to the personnel in a security setup requiring confidentiality. Thus, these *antimagic* graphs could serve as a security model for various kinds of buildings in a scenario of urban planning

In *double wheel* and *helm* (**Figures 7** and **10**) of Refs. [12, 13], the central *antimagic* vertex could serve as the central headquarters for all the offices or the vertices and their passwords/codes or edges being the *antimagic* labeled edges. The *antimagic centerless wheels* $CW2_n$, $CW3_n$, and

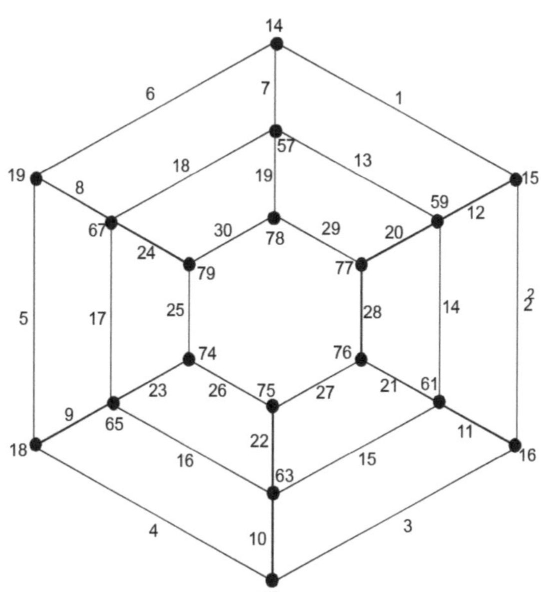

Figure 9. Antimagic centerless wheel-3 $CW3_6$.

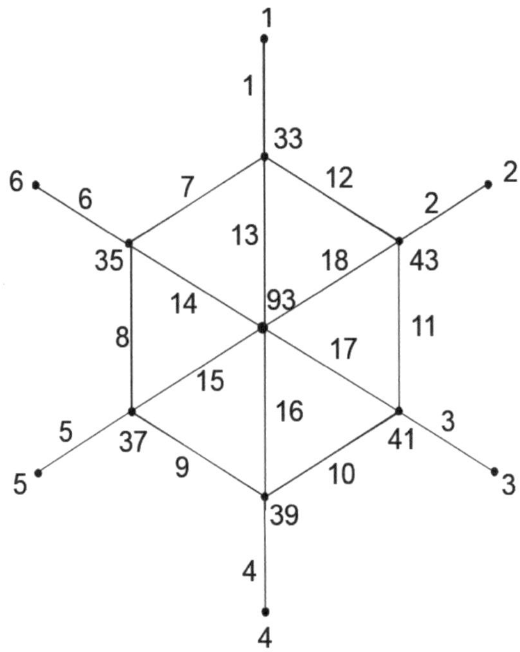

Figure 10. Antimagic helm.

regular actinia of Refs. [12, 13] in **Figures 8**, **9**, and **11** could serve as model for a security system without centralized control. The concepts of **cut-set S** in graph G (removal of all the edges in S disconnects G; removal of some but not all of the edges in S, does not disconnect G) and **vertex cut-set S_1** (analogous for vertices) could be used to disallow certain security personnel's

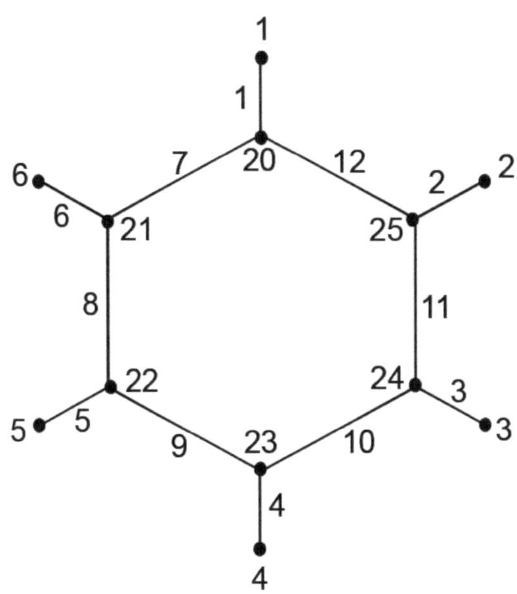

Figure 11. Antimagic regular actinia.

http://dx.doi.org/10.5772/intechopen.68690

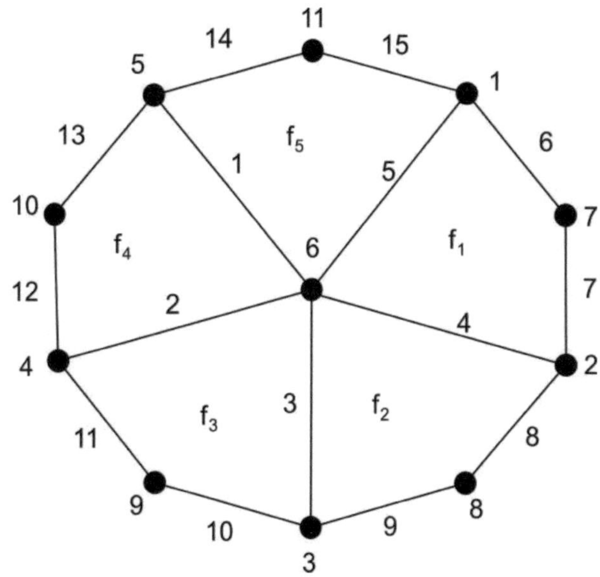

Figure 12. Inner antimagic flower-1. Inner antimagic inner antimagic labels and internal face labels are as follows: (1) 39, 45, 51, 57, 63. $f_1 = 1, f_2 = 2, f_3 = 3, f_4 = 4, f_5 = 5$. (2) 43, 47, 51, 55, 59. $f_1 = 5, f_2 = 4, f_3 = 3, f_4 = 2, f_5 = 1$.

legitimate access. The ideas of **vertex connectivity** (minimum number of vertices deleted to disconnect the graph) and **edge connectivity** (analogous for edges) could also be combined with these applications to check for the violation of codes/passwords.

The graphs studied in Ref. [4] are *wheels, flower-1, and flower-2* (**Figures 6**, **12**, and **13**). The planar graph *flower-1* has one central vertex and rest being outer vertices, and all the internal

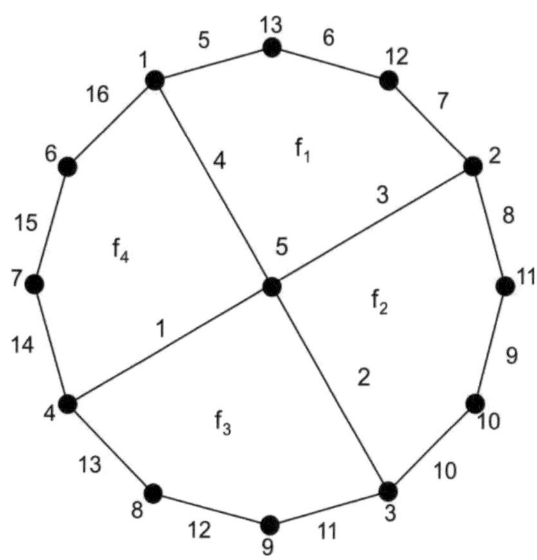

Figure 13. Inner Antimagic Flower-2. Inner antimagic inner antimagic labels and infernal face labels are as follows :(1) 59, 65, 71, 77; f_1 =1, f_2 =2, f_3 =3, f_4 =4. (2) 62, 66, 70, 74; f_1 =4, f_2 =3, f_3 =2, f_4 =1.

faces are bound by four edges. *Flower-2* is a planar graph with one central vertex and rest being outer vertices and all internal vertices are bound by five edges. *Wheels* are found to have *inner magic* as well as *inner antimagic* labelings and *flower-1* and *flower-2* have *inner antimagic* labelings.

The larger of these graphs may have to be checked for existence of these labels. These graphs could also serve as models for surveillance or security systems, in designing building in an urban planning setup, network addressing, communication studies, etc.

Inner magic wheel could be used in a security system where matching of all the labels to the *inner magic* label could lead to decode or unlock the security system.

Above mentioned *antimagic, inner magic,* and *inner antimagic* graphs could have applications in cryptography also.

Ref. [14] mentions the need for efficient methods of graph labelings. Such algorithms are found in Ref. [9] for *graceful, harmonious, sequential, felicitous, and antimagic labelings* for ***bipartite trees where edges do not cross*** useful in communication networks and circuit design where the wires are such that there are **no crossing points**. The various algorithms developed for *bipartite trees,* where edges do not cross with p vertices and q edges for *graceful, harmonious, sequential, felicitous, and antimagic labelings,* could be explored further for applications where such *bipartite trees* serve as models.

5. Results in graph labelings using computer software

The software developed in Refs. [15, 16] to check the existence of major graph labeling methods namely *harmonious, sequential, graceful, felicitous, antimagic, magic* for an arbitrary graph has been used in certain studies. It was used in the studies done in Refs. [3, 17]. Statistical analysis of the antimagic graph labelings of *paths, cycles, wheels,* and *star* graphs in Ref. [3] shows that there is a relation between *degree* (*degree* is the number of edges incident on a vertex) of vertices and number of *antimagic* labelings obtained for these graphs. Presence of higher *degree* vertices increases the number of *antimagic* labelings obtained by the computer software. Moreover, various kinds of graphs were tested, and nearly all the graphs had *antimagic* labelings.

In the application of *antimagic* labelings to *complete graphs* (see **Figure 14**; in a *complete graph,* every vertex is connected to every other vertex), higher *degree* vertices imply more number of legitimate routes for persons who are allowed access to a particular office (where vertices represent offices of the graph and *antimagic* label for a vertex can represent the code/password). *Star* (see **Figure 15**) is a graph with a central vertex and the rest are *pendant vertices* (vertices of *degree* one). *Antimagic Stars* could represent one centralized control room with scattered single offices as the *pendant vertices. Antimagic Paths* could represent linear offices; for example, along a street.

Ref. [18] studies the effect of repeated vertex labels and shows that the labelings produced are faster and several in number on the computer when a vertex label is repeated. This also yields

http://dx.doi.org/10.5772/intechopen.68690

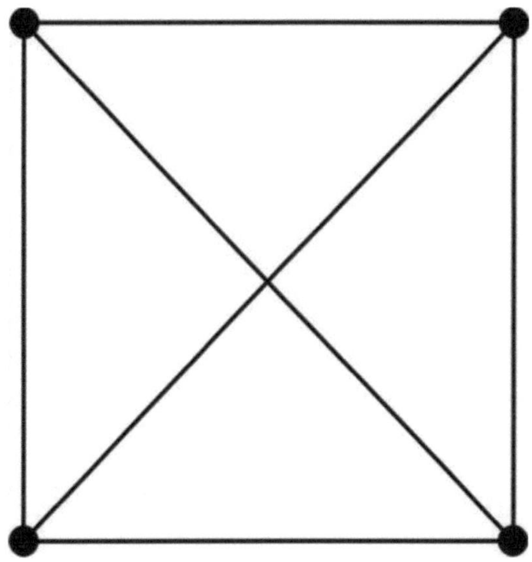

Figure 14. Complete graph.

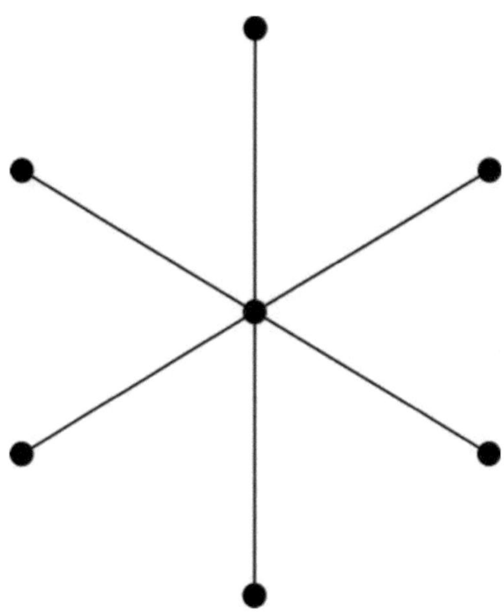

Figure 15. Star.

labeling where there is no labeling with the conventional definition. The repetition of a vertex label can have significance in applications where the repeated vertex label could represent some quantity which is repeated. This opens up possibilities for studies to obtain more labelings and can be explored further.

In Ref. [19], certain patterns in graph labelings have been discovered. It has been discovered that lower degree vertex can be manipulated more easily to yield labeling faster with the help of the software developed in Refs. [15, 16]. This approach could be followed while working manually also.

6. Lexicographic order in graph labelings

Applications so far undiscovered could be explored in the graph labelings in the following works which could be of further interest in research: *lexicographic order* of permutations and combinations is mentioned in Ref. [20] and made use of it to develop a generalized software to check existence of major graph labeling schemes namely *graceful, harmonious, felicitous, sequential, magic, and antimagic* for an arbitrary graph in Refs. [15, 16]. This software has been used to prove that harmonious graph is NP-Complete as given in Ref. [21]. Existence of *lexicographic order* in graph labelings has been studied in Refs. [22, 23].

In a *lexicographic order,* the word 'card' comes before 'cart' as in a dictionary or lexicon.

Permutations in *lexicographic order* of 4! = 24 permutations which are as follows: 1234, 1243, 1324, 1342, 1423, 1432, 2134, 2143, 2314, 2341, 2413, 2431, 3124, 3142, 3214, 3241, 3412, 3421, 4123, 4132, 4213, 4231, 4312, 4321.

Combinations in *lexicographic order* of ${}^6C_4$ are: 1234, 1235, 1236, 1245, 1246, 1345, 1346, 1356, 1456, 2345, 2356, 3456.

In *paths,* permutation 1 and permutation 2 give the *antimagic* labeling for *even* and *odd edges,* respectively, in [23] and is shown in **Figure 16**.

Let us look at the computer generated results for the *cycles* where the edges e_1, e_2, e_3…,and so on are labeled in the order of the permutations in Refs. [17] and is shown in **Figure 17**.

C_4: Permutation 2: 1 2 4 3

Vertex labels (in ascending order): 3 4 6 7

C_6: Permutation 2: 1 2 3 4 6 5

Vertex labels (in ascending order): 3 5 6 7 10 11

C_8: Permutation 2: 1 2 3 4 5 6 8 7

Vertex labels (in ascending order): 3 5 7 8 9 11 14 15

C_{10}: Permutation 2: 1 2 3 4 5 6 7 8 10 9

Vertex labels (in ascending order): 3 5 7 9 10 11 13 15 18 19

Now, let us look at the odd Cycles:

C_5: Permutation 1: 1 2 3 4 5

Vertex labels (in ascending order): 3 5 6 7 9

C_7: Permutation 1: 1 2 3 4 5 6 7

Vertex labels (in ascending order): 3 5 7 8 9 11 13

Permutation 3: 1 2 3 4 6 5 7

Vertex labels (in ascending order): 3 5 7 8 10 11 12

C_9: Permutation 1: 1 2 3 4 5 6 7 8 9

Vertex labels (in ascending order): 3 5 7 9 10 11 13 15 17

Permutation 3: 1 2 3 4 5 6 8 7 9

Vertex labels (in ascending order): 3 5 7 9 10 11 14 15 16

C_{11}: Permutation 1: 1 2 3 4 5 6 7 8 9 10 11

Vertex labels (in ascending order): 3 5 7 9 11 12 13 15 17 19 21

Permutation 3: 1 2 3 4 5 6 7 8 10 9 11

Vertex labels (in ascending order): 3 5 7 9 11 12 13 15 18 19 20

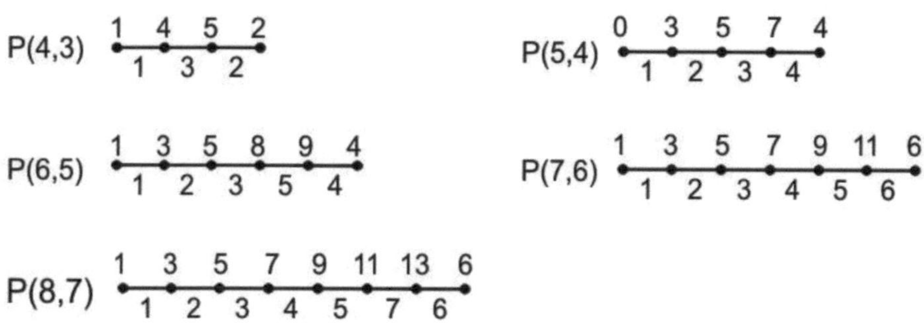

Figure 16. Antimagic paths.

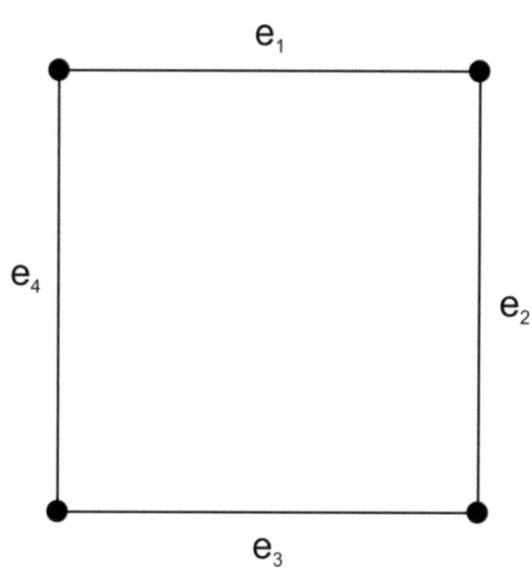

Figure 17. Cycle.

In *cycles*, permutation 1 gives *antimagic* labeling for *odd cycles* (*odd number of edges*), whereas permutation 2 gives *antimagic* labeling for *even cycle* (*even number of edges*). Permutation 3 also gives *antimagic* labeling for *odd cycles* except C_5. Permutation 1 results in induced vertex labels showing almost an *arithmetic progression* with difference 2.

Lexicographic order found in graph labelings could be studied for further applications.

7. Conclusion

Thus, we have seen that the various kinds of delusions and hallucinations often considered "unreal" of the severe psychological disorder of *paranoid schizophrenia*, a kind of *schizophrenia* have much truth and underlying inherent causes in them with insights provided in these factors in this work. Hallucinations and delusions exist in real terms have been proved by *mathematical logic* also.

From the viewpoint of things appearing in creation as dualities of nature, we can see that every pair of opposites like pleasure-pain, right-wrong, day-night, birth-death, stress-relief, for-against, etc. have truth in both the opposites. Thus, the "real"-"unreal" of hallucinations and delusions also have truth in them depending upon the viewpoint of the subject: patient and doctor, respectively. The duality of brain(logical)-heart(feeling) also brings about an unbalanced personality in an individual when there is an overemphasis on calculative logic in human society and overriding the voice of the heart/conscience. Nature brings evolution by these dualities in creation leading to an increase in understanding and knowledge of some unresolved problem.

The *graceful tree conjecture* has elements of psychological conditioning in it, being believed to be true even without a proof for about five decades. Conditioned by a long history of being used to the definitions in traditional Mathematics as the controlling factor in predicting behavior of *all* the defined mathematical structures will lead to projecting the same in the definition of *tree* also. The big difference in the definitions of traditional branches of Mathematics and a newer subject like Graph Theory is that the definition of tree is *unquantifiable and pictorial* thus *cannot* predict the behavior of *all* trees. A *tree* with four vertices is defined as "connected and acyclic" and a *tree* of two million vertices is also defined in the same way with no restrictions and *no measure* giving rise to infinite shapes for larger trees. In the case of a number being divisible by 2, 3, or 5, the definition is applicable to *any number no matter how large* thus *all* such numbers are predicted by this quantifiable definition unlike the *unquantified, pictorial representation* of a *tree*.

Applications of *antimagic, inner magic, and inner antimagic* graphs in urban planning and security setup, etc. have been given for various kinds of graphs. More applications for *bipartite trees where edges do not cross* for various kinds of labelings could be discovered where such trees with *graceful, harmonious, felicitous, sequential, and antimagic* labeling could serve as models. Some results of graph labelings using computer software developed to check the existence of major graph labeling schemes for an arbitrary graph have been discussed. Role of repeated vertex labels and low degree vertex has been presented. Finally, for further studies, the *lexicographic order* found in *antimagic* labelings of paths, cycles, etc. could be studied for discovering more applications.

Author details

Auparajita Krishnaa

Address all correspondence to: akrishnaa1@gmail.com

Department of Mathematics and Statistics, Mohan Lal Sukhadia University, Udaipur, Rajasthan, India

References

[1] Krishnaa A. A note on some thoughts on the graceful tree conjecture. Journal of Discrete Mathematical Sciences and Cryptography. 2013;**16**(6):387-392. DOI: 10.1080/09720529.2013.858484

[2] Krishnaa A. Some applications of labelled graphs. International Journal of Mathematics Trends and Technology. 2016;**37**(3):19-23

[3] Krishnaa A, Dulawat MS, Rathore GS. A study on statistical analysis of antimagic graph labeling. Journal of Rajasthan Academy of Physical Sciences. 2004;**3**(2):111-120

[4] Krishnaa A, Dulawat MS. Algorithms for inner magic and inner antimagic labelings for some planar graphs. Informatica (Lithuania). 2006;**17**(3):393-406. DOI: 10.1016/s0012-365x(01)00175-3

[5] Ahuja N. A Short Textbook of Psychiatry. 7th ed. Jaypee Brothers Publishers, New Delhi (India), St. Louis (USA), London (UK)

[6] Puri BK, Treasadan IH. Textbook of Psychiatry. 3rd ed. Churchill Livingstone Elsevier, UK.

[7] Feldman RS. Essentials of Understanding Psychology. 10th ed. McGraw Hill, New York, USA

[8] Grimaldi RP. Discrete and Combinatorial Mathematics: An Applied Introduction. 4th ed. Pearson Education, USA

[9] Krishnaa A. A study of the major graph labelings of trees. Informatica. 2004;**15**(4):515-524. https://pdfs.semanticscholar.org/16a2/34ed8def26de0eea444907e891970db1f7e9.pdf

[10] Kraayenbrink N, de Frits N, Vavic M. Symmetries in Graceful Trees. www.st.ewi.tudelft.nl/sat/reports/SymGracefulTrees.pdf

[11] Krishnaa A. A note on perception of definitions in graph labelings of trees. Journal of Computer and Mathematical Sciences. 2011;**2**(2):254-259

[12] Krishnaa A. On antimagic labellings of some cycle related graphs. Journal of Discrete Mathematical Sciences and Cryptography. 2012;**15**(4&5):225-235. www.tandfonline.com/doi/pdf/10.09720529.2012.10698377

[13] Krishnaa A. Formulas and algorithms for antimagic labelings of some helm related graphs. Journal of Discrete Mathematical Sciences and Cryptography. 2016;**19**(2):435-445. DOI: 10.1080/09720529.2015.1130935

[14] Daykin JW, Iliopoulos CS, Miller M, Phanalasy O. Antimagicness of generalized corona and snowflake graphs. Mathematics in Computer Science. 2015;**9**(1):105-111

[15] Krishnaa A. Computer modelling of graph labelings. In: Proceedings of National Conference on Mathematical and Computational Models; Coimbatore, India. Allied Publishers: India; 2001

[16] Krishnaa A, Dulawat MS. Algorithms and programs for some problems in graph labellings, operations research and combinatorics. Journal of Mathematics and System Sciences. 2007;**3**(1):1-15

[17] Krishnaa A. On the use of computers in graph labeling. International Journal of Computer Science and Communication. 2012;**3**(1):191-197

[18] Krishnaa A, Dulawat MS. Study of the effect of repeated vertex labels. Ultra Scientist of Physical Sciences. 2011;**23**(1):176-180

[19] Krishnaa A, Dulawat MS, Rathore GS. Some patterns in graph labelings. Pure and Applied Mathematika Sciences. 2005;**LXI**(1-2):61-65

[20] Liu CL. Discrete Mathematics. Tata McGraw Hill. New Delhi, India; 1978

[21] Krishnaa A. A note on a graph scheme and its computational complexity. Journal of Computer and Mathematical Sciences. 2011;**2**(3):521-524

[22] Krishnaa A, Dulawat MS. Lexicographic ordering in graph labelings of cycles, paths and complete bipartite graphs. South East Asian Journal of Mathematics and Mathematical Sciences. 2009;**7**(2):87-93

[23] Krishnaa A, Dulawat MS. Labeling of paths. Journal of Rajasthan Academy of Physical Sciences. 2006;**5**(1):99-104

Chapter 3

Graph-Based Decision Making in Industry

Izabela Kutschenreiter-Praszkiewicz

Additional information is available at the end of the chapter

http://dx.doi.org/10.5772/intechopen.72145

Abstract

Decision-making in industry can be focused on different types of problems. Classification and prediction of decision problems can be solved with the use of a decision tree, which is a graph-based method of machine learning. In the presented approach, attribute-value system and quality function deployment (QFD) were used for decision problem analysis and training dataset preparation. A decision tree was applied for generating decision rules.

Keywords: decision tree, decision-making, machine learning, quality function deployment (QFD), inquiry planning

1. Introduction: decision-making in industry

The decision-making process in industry is focused on finding answers for the following questions: what should be done, how it should be done, when and who by? The decision-making process often uses heuristic and expert knowledge, which respects the relations between different variables, e.g. the problem of inquiry planning needs analysis, such as response time to consumer inquiries, response preparation costs and the risk related to the manufacturing process [1, 2]. There are different methods which aid industrial data analysis, among which quality function deployment (QFD) turns out to be useful in data analysis related to customer inquiries.

Each decision type requires data and process analysis. Decision problems can be divided into categories, distinguished from different points of view. In industry, we can meet structured decisions, as well as unstructured decisions, in which each decision-maker can use different data and processes to reach the conclusion, and semi-structured decisions in which decision scenarios have some structured and unstructured components. Decision problems are caused by a change related to distinctive features (attributes).

Decision-making in industry can be focused on different time periods, i.e. strategic decisions concern a few years, and tactical decisions relate to a period of a few months, whereas operational decisions regard a few days [3].

The decision-making process starts from decision problem analysis [4]. Steps in decision-making model include [5]:

- definition of the problem,
- establishment or enumeration of all the criteria (constraints),
- consideration or collection of all the alternatives,
- identification of the best alternative,
- development and implementation of an action plan,
- evaluation and monitoring of the solution and feedback examination, if necessary.

A problem should be precisely identified and described. Manufacturing products in an industrial plant require combined and coordinated efforts of people, machinery and equipment [6] which create a manufacturing system. This manufacturing system needs suitable values of decision variables, which characterise product and whole stages of the manufacturing process.

The overall manufacturing system decision problems include, among others [2, 6]:

- inquiry planning,
- the problem of resource requirements,
- the problem of resource layout,
- the problem of material flow,
- the problem of buffer capacity.

Decision-making in manufacturing systems can be characterised as follows [6, 7]:

- Manufacturing systems should be able to produce products according to customer requirements.
- Manufacturing systems consist of many interacted components.
- Manufacturing system is changing in time periods.
- Manufacturing systems are influenced by internal and external variables.
- A manufacturing system is complicated, and it is difficult to create its complex model. The relations between variables, which describe it, usually cannot be expressed analytically.
- Data charactering a manufacturing process may be difficult to measure.
- Decisions in a manufacturing process can be focused on achieving different goals, which are sometimes in conflict.

The methods useful in supporting decision-making in manufacturing system include:

- Mathematical programming (linear programming) useful for decision problems for which it is possible to formulate goals and constrains as equations.
- The queuing theory, which is a study of behaviour of queueing systems through the formulation of analytical models [6]. Queue disciplines include FIFO (first in, first out), LIFO (last in, first out), SIRO (service in random order), PRI (priority ordering) and GD (any other specialised ordering).
- Artificial intelligence, including, among others, decision tree and rule-based systems, neural networks and genetic algorithms.
- Simulations, which in industrial applications include the following steps: formulating the problem, collecting data and defining a model, model statistics of system randomness, ensuring validity, constructing and verifying a computer model, pilot runs and validity checks, design experiments, performing runs, analysing output data, documenting and implementing results [6].

In the presented approach, a decision problem should be described with the use of an attribute-value system, which is one of the well-known models of knowledge representation. Under the object-attribute-value (O-A-V) scheme, an object is associated with various attributes, and each attribute is assigned with appropriate values [8]. The attribute-value system uses statement object-attribute-value for decision problem characteristics, e.g. a decision problem like machine tool selection can be characterised by different objects, such as machine, material, process, etc. Each object can be characterised by different attributes (variables), e.g. machine can be characterised by attributes such as type, technical condition, work parameters, etc. Attributes can be characterised by a categorised, numerical or linguistic value.

Decisions can be supported by different types of systems, such as (**Table 1**) [9, 10]:

- transaction processing systems (TPS), which focus on data evidence,
- decision support systems (DSS), which support decision-making using simulation and data processing applicable for different variants,
- expert systems (ES), which support experts in their decisions using heuristic knowledge.

Decision type	**Decision time period**			**Support system**
	Operations level	**Tactical level**	**Strategic level**	
Structured	Resource planning Delivery registration	Economic analysis	Finance of investments Warehouse localisation	TPS, DSS
Semi-structured	Technical production preparation	Credit assessment Scheduling	Product development planning Quality control	DSS, ES
Unstructured	Software purchase	Recruitment of managers	Technology development	ES

Table 1. Decision problem types.

Among methods useful in structured decision-making, decision trees are widely discussed.

Decision-making process based on graph theory can be based on the following stages:

- Formulating the problem.
- Determination of a set of attributes which characterise the decision problem (for this purpose a QFD matrix can be used), e.g. machine failure diagnosis decision problem can be characterised by attributes such as noise level, vibration, type of failure, etc.
- For each attribute a set of possible values is defined, e.g. the attribute noise level can be characterised by interval numerical values such as <85 dB and >85 dB.
- Finding examples from the past with a solution of the problem and creating a training set, e.g. noise level >85 dB, vibration high, solution = repair Section A.
- Creating a decision tree.
- Solving the new problem with the use of a decision tree.

Decision-making process supported by the machine learning method uses knowledge (experience) which comes from different sources (different experts). In traditional decision-making, an expert develops his knowledge based on his own experience.

Human decision-making can be described as follows:

- Expert no. 1—own experience—decision for a new case based on individual experience
- Expert no. 2—own experience—decision for a new case based on individual experience
- etc.

Graph-based machine learning decision-making can be described as follows:

- Expert no. 1—own experience
- Expert no. 2—own experience
- etc.
- Training set—common experience (set of all known cases)
- Decision tree induction
- Decision for a new case supported by a decision tree

Graph-based decision-making can be compared with neural network. In both cases, the knowledge saved in training set joins experience from different sources, but decision tree can be used for decision tree induction, which exhibits the knowledge in a clear way, instead of a neural network, which is able to predict the sought value without any explanations.

2. Decision tree

A decision tree is a graph which can be used as a model of a categorical variable. A decision tree aims at predicting a categorical (numerical or linguistic) output variable from a set of numerical or linguistic input variables [11]. Decision trees are useful in solving classification and prediction problems [12, 13]. The structure of a decision tree involves a root node, internal nodes, leaf nodes and edges which joint nodes, also called branches (**Figure 1**) [14].

A root node is an initial decision node which includes the main attribute in the decision process. Internal nodes include other attributes which are input variables in the decision process, whereas leaf nodes represent output variables including possible decisions in the decision process. Edges represent categorical values assigned to the attribute. The tree always starts from the root node and grows down by splitting the data at each level into new nodes [14].

Decision trees are one of the machine learning methods. Constructing a decision tree requires a set of decision problem-solving examples which create a training set.

Machine learning from examples and its generalisation ability were discussed, e.g. by Shiue et al. (**Figure 2**) [15]. In machine learning methods, one of the most important tasks is to create a training set of examples. For that purpose, it is necessary to define attributes and their values which are important variables in a given decision problem.

Data in the training set can come from a real manufacturing system or from simulation experiments. Examples of training data are presented in **Table 2**.

In machine learning methods, attributes come from decision problem characteristics. QFD can support attribute selection in industry decision-making. In real applications, attributes' value can use numerical as well as linguistic values (numerical value can be an integer or not integer put as a separate value or in intervals). The relevance of attributes in the decision process can be evaluated with the use of, e.g. Shannon entropy in ID3 algorithm, which is not discussed in this chapter.

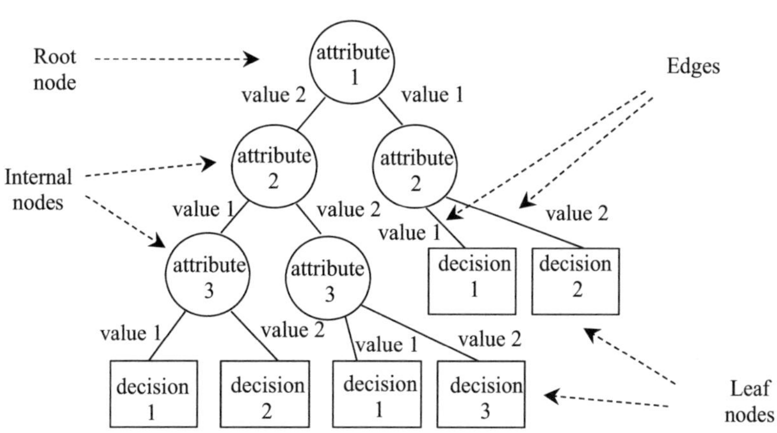

Figure 1. Decision tree.

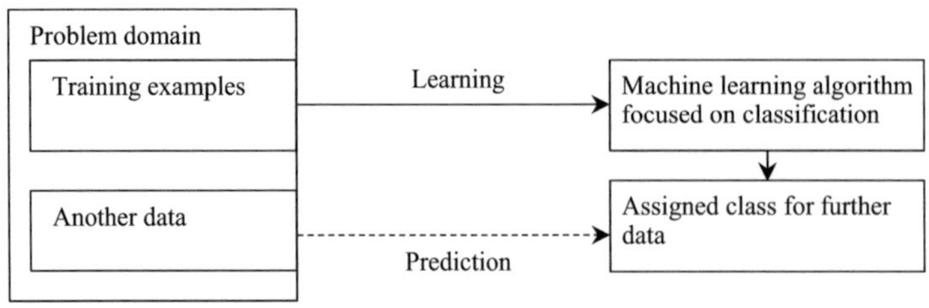

Figure 2. Machine learning from examples.

Case no.	Attribute 1	Attribute 2	Attribute 3	Attribute 4	Decision
	Possible values	Possible values	Possible values	Possible values	Possible values
	(0,1)	(0,1)	(0,1)	(0,1)	(1,2,3)
1	0	1	0	1	1
2	1	1	0	0	2
3	1	1	1	1	2
4	1	0	1	0	1
5	1	0	1	0	1
6	1	0	0	0	2
7	0	0	1	1	3
8	0	0	0	1	3
9	1	1	0	1	1

Table 2. A training set.

Constructing a decision tree classifier is usually divided into two steps: generation and pruning trees (C4.5 and CART algorithms) [16–19].

In the generation phase, the initial tree is built using available training dataset until each leaf becomes homogeneous. In the pruning phase, the already-grown tree is reduced in order to improve the accuracy obtained on the testing dataset [20]. There are many methods for constructing a decision tree. The basic generation algorithm includes the following steps [21]:

1. Start with a single node representing all records in the dataset.
2. Choose one attribute, and split the records according to their values on that attribute.
3. Repeat the splitting on all new nodes, until a stop criterion is satisfied.

A single node representing all records in a dataset was found, and the root attribute was assigned (**Figure 3**).

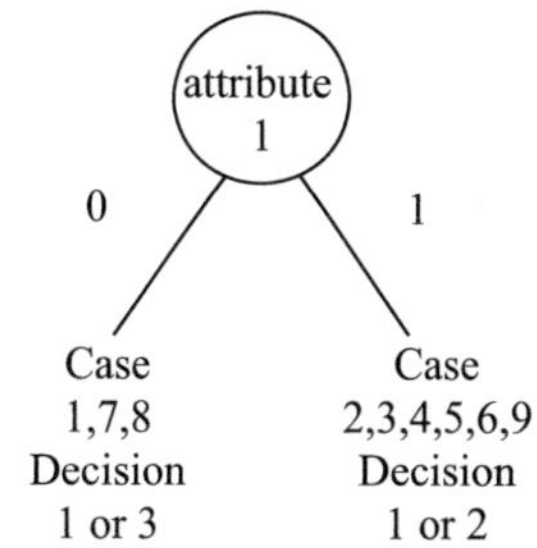

Figure 3. The first node in the decision tree induction process.

In the second step, decisions are not unique, so the decision tree should grow, and other attributes should be taken into consideration (**Figure 4**). In the presented example, the next step is necessary (**Figure 5**). The final decision tree is presented in **Figure 6**.

The decision tree algorithm is a predictive model with a hierarchical structure and used in data mining [22]. This algorithm has several advantages [16, 23–25]:

- The training set can include expert knowledge, as well as results of experiments and industrial data which come from manufacturing processes.
- Decision trees can handle both linguistic and numerical input and output variables (attributes).
- Results of the prediction process are easy to interpret, clear and close to human reasoning.
- It can be joined with other algorithms.
- It is possible to use decision trees even if datasets have missing values.

Decision tree construction requires data preprocessing.

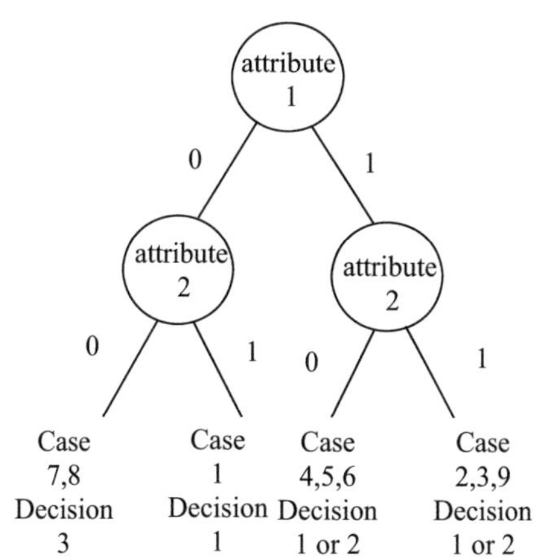

Figure 4. Sequent nodes in the decision tree induction process.

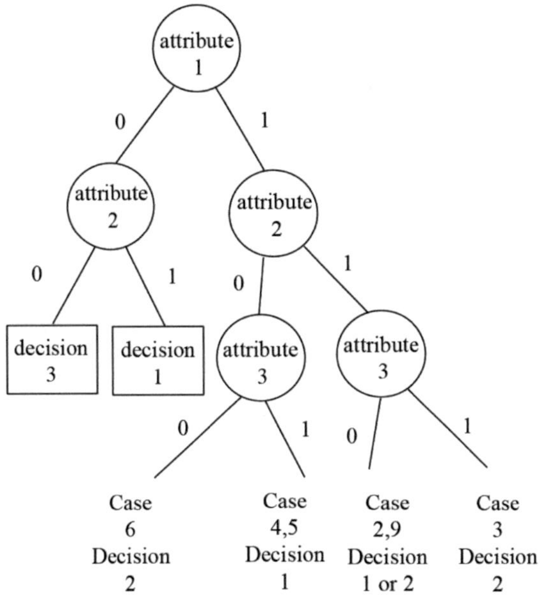

Figure 5. Sequent nodes in the decision tree induction process.

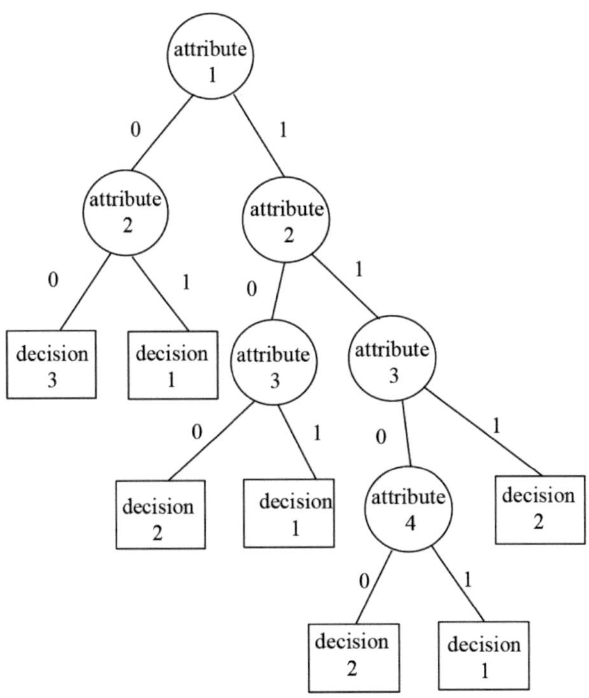

Figure 6. Sequent nodes in the decision tree induction process.

http://dx.doi.org/10.5772/intechopen.72145

3. Data preprocessing in the inquiry planning problem with the use of QFD

Constructing a training set, it is necessary to collect data related to the decision process. Data preprocessing related to inquiry planning can be supported with the use of an attributed model of the product [26, 27].

In the attributed product model, the product functions can be characterised by a set of attributes:

$$F = \{f_1, f_2, \ldots, f_n\} \quad (1)$$

Based on a toothed gear example, the set of attributes includes: f_1—reducer working arrangement; f_2—kind of duty.

Each attribute takes a value from the set F_n^w:

$$F_n^w = \{f_{n1}^w, f_{n2}^w, \ldots, f_{nl}^w\} \quad (2)$$

An example of a set of attribute values F_n^w includes: f_{11}^w—parallel axes; f_{12}^w—perpendicular axes; f_{21}^w—light duty; f_{22}^w—medium duty; f_{23}^w—heavy duty.

The set of product types was denoted as P:

$$P = \{p_1, p_2, \ldots, p_m\} \quad (3)$$

An example of a set of products includes: p_1—helical gear; p_2—bevel-helical gear.

Each product-type p_m includes products $p_{m1}, p_{m2}, \ldots, p_{mk}$, described by attributes p_{mkz}:

$$P_{mk} = \{p_{mk1}, p_{mk2}, \ldots, p_{mkz}\} \quad (4)$$

An example of a set of products includes: p_{11}—one-stage helical-geared reducer mounted on the feet; p_{12}—one-stage helical-geared reducer hanged on the shaft.

An example of a set of product attributes includes: p_{111}—weight; p_{112}—dimensions.

Each attribute p_{mkz} takes a value from the P_{mkz}^w set:

$$P_{mkz}^w = \{p_{mkz1}^w, p_{mkz2}^w, \ldots, p_{mkzt}^w\} \quad (5)$$

Product p_m consists of modules/elements belonging to the set M:

$$M = \{m_1, m_2, \ldots, m_k\} \quad (6)$$

Each element is described by attributes belonging to the set M_k:

$$M_k = \{m_{k1}, m_{k2}, \ldots, m_{kv}\} \tag{7}$$

Examples of attributes describing product parts are: m_{k1}—weight; m_{k2}—type of material.

Attribute values belong to the set M_{kv}^{w}:

$$M_{kv}^{w} = \{ m_{kv1}^{w}, m_{kv2}^{w}, \ldots, m_{kvg}^{w} \} \tag{8}$$

M_k^* is a set of variants of element m_k:

$$M_k^* = \{m_{k1}^*, m_{k2}^*, \ldots, m_{kl}^*\} \tag{9}$$

M_{kv}^{w} is a set of attribute values:

$$M_{kv}^{w} = \{m_{kv1}^{w}, m_{kv2}^{w}, \ldots, m_{kvg}^{w}\} \tag{10}$$

Basing on the presented product-attributed model, it is possible to analyse product and process attributes. The theory helpful in complex product and process development is quality function deployment (QFD), also known as house of quality. QFD supports meeting customer requirements in product and process design (**Figure 7**). QFD is a method of data analysis related to customer requirements, product and process characteristics in industrial plants.

The QFD matrix (**Figure 8**) [28] determines the relations between customer needs (denoted as 'what's') and design characteristics (denoted as 'how's'). The top part of the matrix called a 'roof' indicates how design characteristics interact. The right part of the matrix includes assessment of alternative products. The characteristic of alternative products is presented at the bottom of the matrix. The correlation between 'what's' and 'how's' is registered in the middle part of the matrix.

QFD consists of a series of matrices. The first one represents the relation between customer requirements and product characteristics.

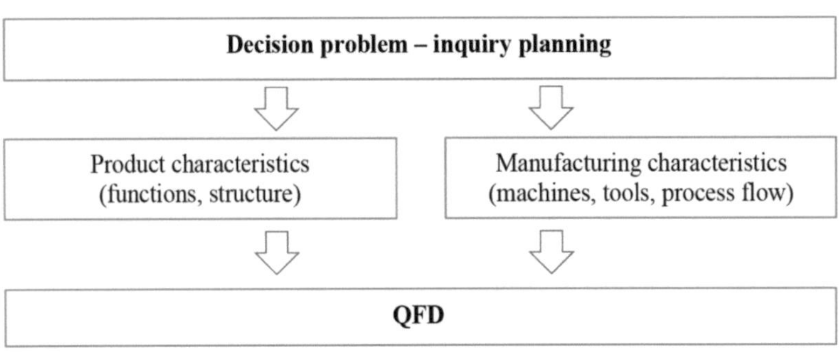

Figure 7. QFD in decision problem solving.

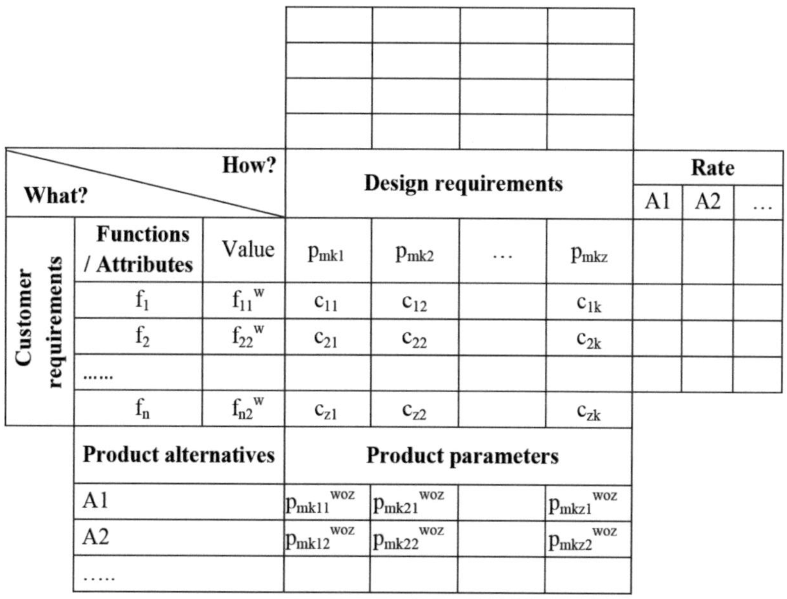

Figure 8. A QFD matrix.

No	Attribute (value 1, value 2, etc.)			Decision
	Height (low, high)	**Colour (dark, red, white)**	**Complication (simple, complex)**	
1	Low	White	Simple	w2
2	High	White	Complex	w1
3	High	Red	Simple	w2
4	Low	Dark	Simple	w1
5	High	Dark	Simple	w1
6	High	White	Simple	w2
7	High	Dark	Complex	w1
8	Low	White	Complex	w1
9	Low	Dark	Simple	w1
10	Low	Red	Simple	w2

Table 3. Training dataset.

The second one describes the relation between product characteristics and product parts.

The third QFD matrix gives information related to the production process.

The fourth one provides information related to production process parameters [28–30].

QFD is a customer-oriented method of industrial data analysis which is able to take into consideration numerical, as well as linguistically specific variables. QFD aids relations between customers and industrial processes.

An example of QFD was developed on the basis of the training dataset presented in **Table 3** (**Figure 9**). The possible values of chosen attributes characterising products from the customer's point of view were specified in the left part of the matrix.

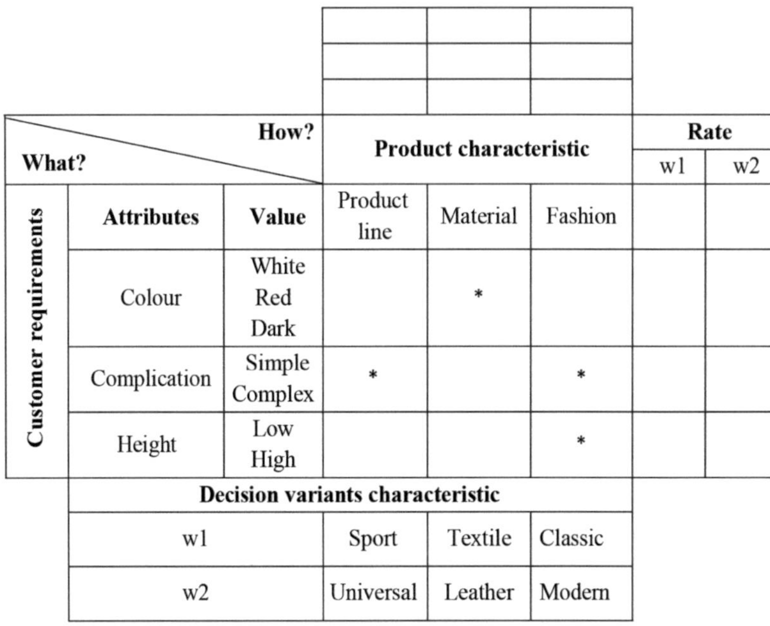

Figure 9. An example of QFD matrix.

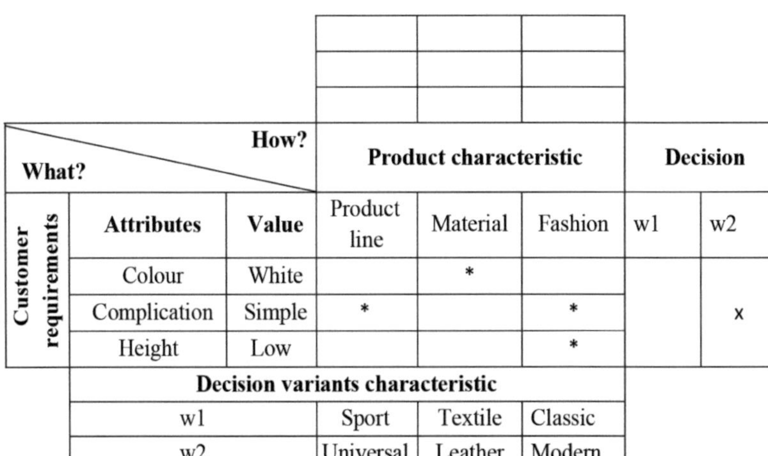

Figure 10. The first client's decision.

What? \ How?			Product characteristic			Decision	
Customer requirements	**Attributes**	**Value**	Product line	Material	Fashion	w1	w2
	Colour	White		*			
	Complication	Complex	*		*	x	
	Height	High			*		
	Decision variants characteristic						
	w1		Sport	Textile	Classic		
	w2		Universal	Leather	Modern		

Figure 11. The second client's decision.

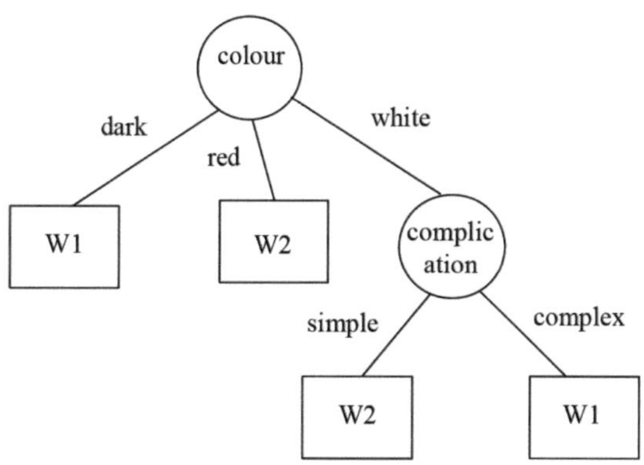

Figure 12. A decision tree example.

The first client's decision is presented in **Figure 10**. The data from this matrix creates the first record in the training dataset presented in **Table 3**.

The second client's decision is presented in **Figure 11**; this data create the second record in training dataset presented in **Table 3**.

Decision tree induction starts from the 'colour' attribute, followed by the 'complication' attribute (**Figure 12**).

4. Knowledge extraction from a decision tree: production rules for knowledge representation

Decision tree induction is closely related to rule induction; each path from the root of a decision tree to one of its leaves can be transformed into a rule [16], which is one of the most popular approaches to knowledge representation. Rules, sometimes called IF-THEN rules, can take various forms, e.g.:

Simple rules:

- IF condition THEN action
- IF premise THEN conclusion

Complex rules:

- IF proposition p1 AND proposition p2 are true THEN proposition p3 is true

Some of the benefits of IF-THEN rules:

- they are modular,
- each rule defining a relatively small and independent piece of knowledge.

For example, paths from the decision tree presented in **Figure 12** can be transformed into rules:

- IF colour = dark THEN w1
- IF colour = red THEN w2
- IF colour = white AND complication = simple THEN w2
- IF colour = white AND complication = complex THEN w1

The resulting set of rules can be transformed to improve its comprehensibility for a human user, and possibly its accuracy [31], e.g. the rules presented above can be transformed to complex rules, such as:

- IF colour = dark OR (colour = white AND complication = complex) THEN w1
- IF colour = red OR (colour = white AND complication = simple) THEN w2

5. The prediction process

Rules produced in Section 4 can be used for prediction. For further clients, for whom requirements are characterised in the left part of the matrix presented in **Figure 13**, in the inquiry planning process, the enterprise should offer product 'w2'.

The decision-making based on the rules produced in Section 4 is presented in **Table 4**.

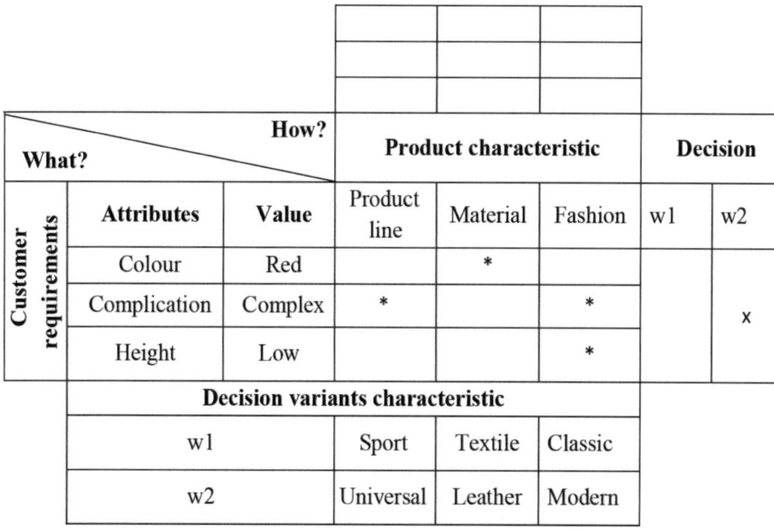

Figure 13. The second client's decision.

No	Attribute (value 1, value 2, etc.)			Decision
	Height (low, high)	**Colour (dark, red, white)**	**Complication (simple, complex)**	
11	Low	Red	Complex	w2

Table 4. Predicted values.

6. Conclusions

A decision tree is one of the graph-based methods of machine learning which can be used in decision making in industry. Among decision problems met in industry, one of the most important decision-making is inquiry planning, which starts from product definition offered to a particular client. QFD can be applied as a method which facilitates data preprocessing in inquiry planning. What is more is that QFD aids attribute specification important from both customer and engineering points of view.

In decision tree induction, a training set can use categorical number values, as well as linguistically specific attributes. In training dataset development, the attribute-value system is a useful method of data analysis. The main steps in decision tree induction were applied. Optimal decision tree induction can be fulfilled with different algorithms which were not discussed.

A decision tree is called the 'white box' method because of clarity and intelligibility for humans, which is important in the decision-making process in industrial context.

The presented approach can be applied in e-commerce systems which are currently under development in many branches of the industry.

Author details

Izabela Kutschenreiter-Praszkiewicz

Address all correspondence to: ipraszkiewicz@ath.bielsko.pl

University of Bielsko-Biała, Bielsko-Biała, Poland

References

[1] Moravcik O, Misut M. Decision Support Systems in Manufacturing Systems Management. In: Tzafestas SG, editor. Computer-Assisted Management and Control of Manufacturing Systems. London: Springer; 1997

[2] Macioł A, Macioł P, Jędrusik S, Lelito J. The new hybrid rule-based tool to evaluate processes in manufacturing. The International Journal of Advanced Manufacturing Technology. 2015;**79**:1733-1745

[3] Mtsniemi T. Operational Decision Making in the Process Industry Multidisciplinary Approach. VTT Technical Research Centre of Finland; 2008

[4] Kepner C, Tregoe B. The New Rational Manager: An Updated Edition for a New World. Updated ed. Princeton, NJ: Princeton Research Press; 1997

[5] Guo K. DECIDE: A decision-making model for more effective decision making by health care managers. The Health Care Manager. 2008;**27**(2):118-127

[6] Chryssolouris G. Manufacturing Systems. New York: Springer Science + Business Media; 1992

[7] Su C-T, Lin C-S. A case study on the application of Fuzzy QFD in TRIZ for service quality improvement. Quality & Quantity. 2008;**42**:563-578

[8] Hong TY, Tsai DH. An integrated expert operation planning system with a feature-based design model. The International Journal of Advanced Manufacturing Technology. 1994;**9**:305-310

[9] Kwaśnicka H. Sztuczna inteligencja i systemy ekspertowe. Rozwój, perspektywy. Wrocław: Wydawnictwo Wyższej Szkoły Bankowości i Finansów; 2005

[10] Mulawka J. Systemy ekspertowe. Warszawa: WNT; 1996

[11] Voisine N, Boullé M, Hue C. A Bayes Evaluation Criterion for Decision Trees. In: Guillet F et al., editors. Advances in Knowledge Discovery and Management, SCI 292. Berlin, Heidelberg: Springer-Verlag; 2010. pp. 21-38

[12] Micheal N. Artificial Intelligence: A Guide to Intelligence Systems. Great Britain: Addison Wesley; 2002

[13] Zhou BH, Xi LF, Cao YS. A beam-search-based algorithm for the tool switching problem on a flexible machine. The International Journal of Advanced Manufacturing Technology. 2005;**25**(9-10):876-882

[14] Kuo Y, Lin K-P. Using neural network and decision tree for machine reliability prediction. The International Journal of Advanced Manufacturing Technology. 2010;**50**:1243-1251

[15] Shiue Y-R, Guh R-S. The optimization of attribute selection indecision tree-based production control systems. The International Journal of Advanced Manufacturing Technology. 2006;**28**:737-746

[16] Rokach L, Maimon O. Decision Trees. Data Mining and Knowledge Discovery Handbook. US: Springer; 2005

[17] Li Y, Guo J-N. Research on Pruning Algorithm of Decision Tree. Henan Science. 2009; **27**(3):320-323

[18] Wang S-S, Sun J-Y, Li L-L. Fault Decision Tree model in the Application of Expert System. Computer Applications. 2005;**25**(s):293-294

[19] Gong Y, Li Y. Motor fault diagnosis based on decision tree-Bayesian network model. In: Jin D, Lin S, editors. Advances in ECWAC, AISC 148. Vol. 1. Berlin, Heidelberg: Springer-Verlag; 2012. pp. 165-170

[20] Gorunescu F. Data Mining: Concepts, Models and Techniques, ISRL 12. Berlin, Heidelberg: Springer-Verlag; 2011. pp. 159-183

[21] Magnani M, Montesi D. Uncertainty in Decision Tree Classifiers. In: Deshpande A, Hunter A, editors. SUM 2010, LNAI. Vol. 6379. 2010. pp. 250-263

[22] Stravinskienė A, Gudas S, Dabrilaite A. Decision tree algorithms: Integration of domain knowledge for data mining. In: Abramowicz W, Domingue J, Węcel K, editors. BIS Workshops, LNBIP, 127. 2012. pp. 13-24

[23] Larose DT. Data Mining Methods and Models. Hoboken: John Wiley & Sons, Inc.; 2006

[24] Phillips J, Buchanan BG. Ontology-guided knowledge discovery in databases. In: Proceedings of the 1st International Conference on Knowledge Capture. 2001. pp. 123-130

[25] Kohavi R, Quinlan JR. Decision-tree discovery. In: Klosgen W, Zytkow JM, editors. Handbook of Data Mining and Knowledge Discovery, ch. 16.1.3. Oxford University Press; 2002. pp. 267-276

[26] Kutschenreiter-Praszkiewicz I. Zastosowanie metod sztucznej inteligencji w planowaniu prac przygotowania produkcji elementów maszyn. Zakopane: Komputerowo Zintegrowane Zarządzanie; 2011

[27] Kutschenreiter-Praszkiewicz I. Integration of product design and manufacturing with the use of artificial intelligent methods. Journal of Machine Engineering. Vol. 11, No. 1-2, 2011 Model Based Manufacturing Operation. In: Jerzy Jędrzejewski editor. Editorial

Institution of the Wroclaw Board of Scientific Technical Societies Federation NOT. s. 46-53. Wrocław, 2011

[28] Kutschenreiter-Praszkiewicz I. Systemy bazujące na wiedzy w technicznym przygotowaniu produkcji części maszyn. Bielsko-Biała: Wydawnictwo Naukowe Akademii Techniczno-Humanistycznej; 2012

[29] Iranmanesh H, Thomson V. Competitive advantage by adjusting design characteristics to satisfy cost targets. International Journal of Production Economics. 2008;**115**:64-71

[30] Raharjo H, Brombacher AC, Xie M. Dealing with subjectivity in early product design phase: A systematic approach to exploit Quality Function Deployment potentials. Computers & Industrial Engineering. 2008;**55**:253-278

[31] Quinlan JR. Simplifying decision trees. International Journal of Man-Machine Studies. 1987;**27**:221-234

Chapter 4

Governance Modeling: Dimensionality and Conjugacy

Pierre Mazzega, Claire Lajaunie and Etienne Fieux

Additional information is available at the end of the chapter

http://dx.doi.org/10.5772/intechopen.71774

Abstract

The Q-analysis governance approach and the use of simplicial complexes—type of hypergraph—allow to introduce the formal concepts of dimension and conjugacy between the network of entities involved in governance (typically organizations) and the networks of those attributes taken into account (e.g. their competences), which offer a specific angle of analysis. The different sources of existing data (e.g. textual corpora) to feed the analysis of governance—environmental in particular—are mentioned, their reliability is briefly discussed and the required pre-processing steps are identified in the perspective of evidence-based analyses. Various indices are constructed and evaluated to characterize the context of governance as a whole, at mesoscale, or locally, i.e. at the level of each of the entities and each of the attributes considered. The analysis of ideal-type stylizing boundary cases provides useful references to the analysis of concrete systems of governance and to the interpretation of their empirically observed properties. The use of this governance modeling approach is illustrated by the analysis of a health-environment governance system in Southeast Asia, in the context of a One Health approach.

Keywords: governance, modeling, simplicial complex, evidence-based analysis, topology, One Health, ideal type, indices

1. Introduction

In April 2010, in the Gulf of Mexico, started the BP Deepwater Horizon oil spill, considered one of the largest marine oil spills in the history of the petroleum industry (estimated to over 600,000 tons of oil released in Gulf of Mexico over 3 months) killing 11 workers and leading to a major environmental disaster. It raised a number of legal issues involving a variety of actors, various levels of decision-making and regulation (from international to local). Presented as *"an important example of multidimensional governance in action"* by Osofsky [1], it led to an attempt by the same author to provide a conceptual model for understanding complex regulatory problems. If the multidimensional aspect of governance is effectively considered as the central challenge in this complex socio-environmental tragedy and has been debated as such as it will be later on, in the case of climate change litigation [2], it is not at all addressed empirically but stay at a very descriptive level. Furthermore, some descriptions of multidimensionality through legal lenses are in contradiction with the mathematical notion of dimension (cf. the paragraph on multidimensionality in Ref. [3]).

Nevertheless, the notion of dimension is fundamental in mathematics and physics and therefore in disciplines using their formal representations (e.g. in ecology or epidemiology modeling). It is declined in various ways, depending on whether it attempts to characterize the space in which interactions are deployed (embedding dimension, local dimension), the geometry of an object (e.g. fractals) or the development of instabilities that work on the evolution of the state of a system (e.g. Lyapunov dimension) [4, 5]. The analysis of governance by political scientists or international relations scholars has made only an extremely limited use[1] of this notion, which, however, is adaptable to the needs of this field of research and is likely to consolidate an empirical, evidence-based approach of governance.

The situation is similar concerning the notion of conjugacy: if a group of organizations is involved in the management of a set of environmental issues, the symmetrical point of view considers that these issues solicit organizations, thus offering another perspective on governance. This kind of duality of approaches is shown in a conjugate relation between two expressions of a formal entity, in this case a simplicial complex, a particular type of hypergraph [9, 10]. Continuing our approach of providing the network governance study with formal tools and the concepts that they provide [11, 12], we apply in this paper the notions of dimension and conjugacy as used in a discrete modeling of governance based on Q-analysis.

This approach proposed in the 1970s by the mathematician Ron Atkin [13, 14] has been used to formalize various problems in social sciences [15–19]. It is now developed in the context of the application of hypergraphs to the analysis of various complex systems (e.g. [19]). The formalism of the simplicial complexes intervenes in a very wide range of applications (for a brief overview, see e.g. [20–22]). A more general survey of data processing using topology methods is found in [23, 24].

The main notions of Q analysis are presented in Section 2. This study aims at illustrating their use by analyzing a network of health-environment institutions and themes in Southeast Asia are presented in Section 2. The concepts of dimension and conjugacy are presented in Section 3. As soon as the dimension of simplices or of paths in the structure is higher than 3, their representation is not readable. For the analysis, we rely on indicators defined in Section 4. Models corresponding to classical general ideal type of governance are then presented in Section 5 and are used as references to analyze empirical governance systems. In Section 6, we discuss the role of generalist organizations (organizations with a large and diverse portfolio of competences) as seen as high-dimensional simplices in a governance system. A

[1]The abundant indexes of the subjects of three relatively recent synthesis books—the Oxford Handbook of Governance [6], the Oxford Handbook of Political Methodology [7] and the Oxford Handbook of International Relations [8]—do not contain the *dimension*, *conjugacy* or *duality* entries.

discussion on the potential use of this approach and on the introduction of the concepts of dimension and conjugacy in governance analysis is proposed in Section 7, and a short conclusion in Section 8.

2. Actors and competencies

We consider a set of organizations with expertise on themes emerging from the analysis of the emergence or re-emergence of infectious diseases in Southeast Asia in a context of environmental change. Epidemiology shows that human health is likely to be affected by a wide variety of pathogens, themselves dependent on their vectors and hosts and on environmental (precipitation and ambient temperature climatology, surface hydrological regime) or socio-ecological dynamics (land cover and land use, biodiversity state and uses, economic exchanges, migration) [25]. In response to the risks of pandemics, the One Health approach [26, 27] promotes simultaneous consideration of the determinants of human health, animal health (domestic animals and wildlife) and environmental health.

This posture leads to considering both public health and environmental themes—such as climate change [28, 29] or the loss of biological diversity [30]—linked by epidemiological dynamics [31], as well as organizations operating from international to regional or local levels in these areas. Health governance in Southeast Asia, a hot spot for the emergence or re-emergence of infectious diseases and biodiversity [32], is also based on political or legal texts (e.g. international conventions [33]), which are themselves integral parts of governance systems [34]. In the One Health perspective, the following health and environmental themes are identified: human health (HH label), animal health (AH), ecosystem health (EH), climate change (CC), land use and land cover (LU), water resources (WR) and risk assessment or risk analysis (RA). The organizations we consider[2] are listed in **Table** 1 with the themes for which they display competencies.

Types are indicated by combining the following initials: I = international; R = regional; O = organization; Ob = observation; N = network; Po = policy; Pr = project or initiative; NG = non-governmental and PF = platform. The labels of themes read as follows: HH = human health; AH = animal health; EH = ecosystem health; CC = climate change; BD = biodiversity; FS = food security; LU = land use and land cover; WR = water resources and RA = risk assessment or risk analysis.

Under the generic term *organization*, we target organizations as such (FAO, WHO), networks (e.g. TROPMED, APEIR, GEOBON) or network of networks of organizations (CORDS), consortia (MBDS), information systems (ARAHIS), fora (FREH) or technical or cooperation platforms (ARAHIS, EVIPNeT). ASEAN2025 [35] outlines the ASEAN policy strategy for collaboration and development in member countries, and as such participates in regional governance, particularly on health-environment issues. All these entities have an institutionalized existence. Regional health-environment governance involves in fact the diversity of

[2]The criteria and methodology used for this choice of organization are described and discussed in Ref. [11].

	L	T	Websites of organizations	Acronym = Subset of Themes
01	GLOBAL	IO	http://www.fao.org/home/en/	FAO = {HH,AH,EH,CC,BD,FS,LU,WR,RA}
02		IO	www.oie.int/	OIE = {AH,RA}
03		IO	www.who.int/	WHO = {HH,CC,RA}
04		IObN	http://geobon.org/	GEOBON = {BD}
05		IO	www.cordsnetwork.org/	CORDS = {HH,AH,EH,RA}
06	ASEAN	RPo	See reference [35]	ASEAN2025 = {HH,CC,BD,FS,LU,WR,RA}
07		RO	https://www.aseanbiodiversity.org/	ACB = {BD,RA}
08		RPr	http://environment.asean.org/	ACEenv = {HH,CC,BD,WR}
09		RN	www.asfnsec.org/	ASFN = {CC,BD,FS,RA}
10		RO	http://aichr.org/	AICHR = {HH,CC}
11		NGO	https://www.aseanlawassociation.org/	ALAWASS = {RA}
12		RPF	http://www.rr-asia.oie.int/	ARAHIS = {AH}
13	SEAMEO	RO	www.seameo.org/	SEAMEO = {HH}
14		RO	https://www.biotrop.org/	BIOTROP = {HH,AH,EH,BD,WR}
15		RO	www.seameo-recfon.org/	RECFON = {HH,FS,RA}
16		RO	http://www.searca.org/	SEARCA = {CC,BD,FS,LU,WR}
17		RN	http://seameotropmednetwork.org/	TROPMED = {HH,RA}
18	Mekong	RN	www.mbdsnet.org/	MBDS = {HH,RA}
19		RPr	http://www.cdcmoh.gov.kh/25-cdc2-project/	CDC-ADB = {HH,AH,CC,RA}
20		RO	http://www.mrcmekong.org/	MRC = {HH,AH,EH,CC,BD,FS,LU,WR,RA}
21		NGO	www.mekonglawcenter.org/	MRLC = {BD}
22		RN	http://cansea.org.vn/	CANSEA = {CC,BD,FS,LU,WR}
23		RPr	https://www.adb.org/projects/40253-012/main	BCI/CeP = {HH,EH,CC,BD,FS,LU,WR,RA}
24		RPr	http://lowermekong.org/	LMI = {HH,BD,FS,WR,RA}
25	Asia-Pacific	RON	http://www.esabii.biodic.go.jp/ap-bon/	APBON = {BD}
26		RPF	http://www.aehin.org/	AeHIN = {HH,RA}
27		RPF	http://www.who.int/evidence/en/	EVIPNetA = {HH}
28		RO	http://apeir.net/	APEIR = {HH,AH,EH,RA}
29		RO	www.pemsea.org/	PEMSEA = {EH,BD,FS,WR}
30		RO	www.asiadhrra.org/	ADHRRA = {CC,BD,FS}
31		RPF	http://www.esabii.biodic.go.jp/	ESABII = {BD}
32		RPr	http://www.wpro.who.int/rfeh/en/	RFEH = {HH,EH,CC,RA}
33		RPr	www.cobsea.org/	COBSEA = {EH,CC,BD,FS,LU,WR}
34		RN	www.aecen.org/	AECEN = {HH,CC,BD, RA}

Table 1. List of organizations (level L and type T in columns 2 and 3) as simplices over the health-environment-related themes.

organizations and political and legal mechanisms that must be taken into account in an empirical approach.

3. Governance structure: dimensions and conjugacy

To present the concepts we are interested in, we work out a small-size case and introduce some notations. We consider a set X of $M = 4$ organizations (with acronyms WHO, SEARCA, LMI, APEIR), a set Y of $N = 6$ themes or issues (with labels HH, AH, EH, CC, LU, RA) and relation $\mathcal{R}$ so that $x_j \mathcal{R} y_k$ means that organization $x_j \in X$ has competency on theme $y_k \in Y$ as indicated by the checked cells of the table in **Plate 1A**. This information is coded in the incidence matrix R with a 1 at the intersection of *jth* line and *kth* column (zero value otherwise; see **Plate 1B**).

Now consider each organization x_j as the set of themes with which it is related, say $\overline{x_j} \approx \{y_k \text{ such as } x_j \mathcal{R} y_k\}$. For example, we have $\overline{APEIR} = \{HH, AH, EH, RA\}$. The organization APEIR can be represented as a regular polyhedron of four linked vertices (the related themes), say as a tetrahedron or 3-simplex (which is 3-dimensional). In the same way, $\overline{WHO} = \{HH, CC, RA\}$ is a 2-simplex (three vertices, triangle, 2-dimensional). $\overline{SEARCA} = \{CC, LU\}$ and $\overline{LMI} = \{HH, RA\}$ are two disjoint 1-simplices (2 linked vertices, line segment, 1-dimensional). Altogether these simplices form the simplicial complex $K_X^{ex}[Y, \mathcal{R}]$, the subscript X indicating that the simplices represent organizations (elements of X) and the superscript *"ex"* standing for *"example."* **Figure 1A** shows that $\overline{LMI}$ is a 1-common face (line segment) of both the $\overline{APEIR}$ tetrahedron and $\overline{WHO}$ triangle. $\overline{WHO}$ and $\overline{SERCA}$ share a 0-face (with a single vertex $\{CC\}$).

In a symmetrical or conjugated way, we can consider each theme as the set of organizations with which it is bound by the inverse relation $\mathcal{R}^{-1}$: $\overline{y_k} \approx \{x_j \text{ such as } y_k \mathcal{R}^{-1} x_j\}$. The conjugate simplicial complex $K_Y^{ex}[X, \mathcal{R}^{-1}]$ is represented in **Figure 1B**. $\overline{LU} = \{SERCA\}$, $\overline{AH} = \{APEIR\}$ and $\overline{EH} = \{APEIR\}$ are 0-simplices, the last two not being distinguishable in this specific context. $\overline{CC}$ is a 1-simplex; $\overline{HH}$ and $\overline{RA}$ are undistinguishable 2-simplices (same triangle). The $M \times M$ symmetric matrix RR^T (with elements a_{jk}; R^T is the transposed matrix of R) convey information on $K_X^{ex}[Y, \mathcal{R}]$: a_{jj} is the number of vertices forming the simplex $\overline{x_j}$; a_{jk} is the number

$\mathcal{R}$	HH	AH	EH	CC	LU	RA
WHO	×			×		×
SEARCA				×	×	
LMI	×					×
APEIR	×	×	×			×

$$R = \begin{bmatrix} 1 & 0 & 0 & 1 & 0 & 1 \\ 0 & 0 & 0 & 1 & 1 & 0 \\ 1 & 0 & 0 & 0 & 0 & 1 \\ 1 & 1 & 1 & 0 & 0 & 1 \end{bmatrix}$$

Plate 1. (Left) Table of the relation between organizations (lines) and themes (columns) and (right) corresponding incidence matrix.

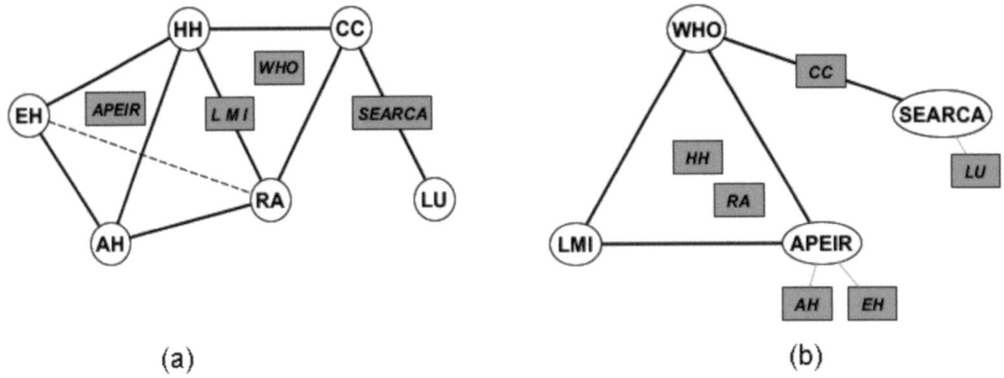

Figure 1. Example of simplicial complex $K_X^{ex}[Y, \mathcal{R}]$ and conjugate complex $K_Y^{ex}[X, \mathcal{R}^{-1}]$. The label of simplices (*resp.* vertices) is given in rectangular gray boxes (*resp.* ellipses). (A) 3D simplicial complex K_X^{ex}. Each simplex is an organization, and the vertices are themes. (B) 2D simplicial complex K_Y^{ex}. Each simplex is a theme, and the vertices are organizations.

of vertices shared by simplices $\overline{x_j}$ and $\overline{x_k}$. In the same way, matrix $R^T R$ encodes information on $K_Y^{ex}[X, \mathcal{R}^{-1}]$.

As for graphs, it is possible to define paths in a simplicial complex, but of various dimensions. The intersection between two simplices—for example $\overline{x_j}$ and $\overline{x_{j+1}}$—is either empty or is a set of vertices that form a simplex $\overline{x_j} \cap \overline{x_{j+1}}$ of K_X^{ex}. Two simplices $\overline{x_1}$ and $\overline{x_m}$ are connected by a path of length $(m-1)$ if the sequence of simplices $\overline{x_1}, \overline{x_2}, \ldots, \overline{x_m}$ satisfies $\overline{x_j} \cap \overline{x_{j+1}} \neq \varnothing$ for every $j \in \{1, 2, \ldots, m-1\}$. It is also a q-path if:

$$\mathrm{Min}_{j=1..(m-1)}\left[\dim\left(\overline{x_j} \cap \overline{x_{j+1}}\right)\right] = q \tag{1}$$

That is to say each pair of consecutive simplices of the sequence shares at least $(q+1)$ vertices. $\overline{x_1}$ and $\overline{x_m}$ are then q-connected. Any path of minimum length between two simplices is called a geodesic. The relation $\overline{x_j}\mathcal{R}_q\overline{x_k}$ if $\overline{x_j}$ and $\overline{x_k}$ are q-connected is an equivalence relation on K_X^{ex}. The equivalence classes of $\mathcal{R}_q$ are called the q-connected components of K_X^{ex}. We denote by Q_q their number. The graphical representation of simplicial complexes is not readable as soon as the dimension of the simplices is greater than 3 or when the complex is composed of numerous simplices with common faces. This limitation is bypassed by the use of indicators.

4. Governance complex: global to local indexes

We define three types of indexes to characterize a simplicial complex K defined from a relation involving a space X of cardinal N. A global index characterizes a simplicial complex in its entirety. A mesoindex takes into account the positioning of each simplex in the whole structure of the simplicial complex. A local index is attached to each simplex and allows to evaluate the configuration of their local insertion, with their immediate neighbors, in the complex. The first global index associated to a simplicial complex K is its dimension *dimK*: it is the dimension of its higher dimensional simplex. In our example of **Figure 1**, we have $dimK_X^{ex} = 3$ and $dimK_Y^{ex} = 2$. The structure vector $Q(K)$ is formed from the number of q-connected components of K, for q varying from 0 to $dimK$:

$$Q(K) = [Q_0, Q_1, Q_2, .., Q_{dimK}] \tag{2}$$

A global size index is evaluated according to the formula:

$$G_{SI}(K) = \frac{2}{(dimK+1)(dimK+2)} \sum_{q=0}^{dimK} (q+1)Q_q \tag{3}$$

If K is complete, then $G_{SI} = 1$. If none of the N vertices of K is connected to another vertex, then $G_{SI} = N$. In order to compare the percentage of dispersion of vertices between complexes that do not have the same number of vertices, one also defines a normalized size index:

$$\overline{G_{SI}}(K) = 100 \times [G_{SI}(K) - 1)]/(N-1) \tag{4}$$

$\overline{G_{SI}}$ can vary from 0% for a clique to 100% for isolated vertices (i.e. for a stable set in the terminology of graph theory). *Meso*-indexes take into account the insertion of each simplex in the network. For each simplex $\overline{x_j}$, one defines a size index by:

$$M_{SI}(\overline{x_j}) = \frac{2}{(N-1)(dimK+1)(dimK+2)} \sum_{q=0}^{dimK} (dimK+1-q) \times n_q(\overline{x_j}) \tag{5}$$

where $n_q(\overline{x_j})$ is the number of simplices $\overline{y}$, with $y \neq x_j$, connected to $\overline{x_j}$ by a q-path. $M_{SI}(\overline{x_j})$ is 0 when $\overline{x_j}$ is isolated (not connected to any other simplex $\overline{y}$, with $\overline{y} \neq \overline{x_j}$). It increases in particular when the dimension of the complex is high and the simplex has connections with many other simplices along low-dimensional q-paths. A path index $P_q(\overline{x_j})$ is also defined for each simplex, which also depends on a threshold dimension q (with q varying from 0 to $dimK$):

$$P_q(\overline{x_j}) = \frac{2}{(N_q-1)\left(g_q+1\right)\left(g_q+2\right)} \sum_{k=1}^{g_q+1} \left(g_q+2-k\right) \times m_{q,k}(\overline{x_j}) \tag{6}$$

where $m_{q,k}(\overline{x_j})$ is the number of simplices $\overline{y}$, with y different from $\overline{x_j}$ and connected by a q-path of length at most k. g_q is the maximum length of q-geodesics and N_q is the number of simplices $\overline{x}$ with dimension greater or equal to q. $P_q(\overline{x_j})$ varies from 0 when $\overline{x_j}$ is isolated (no access to this simplex) to 1 when $\overline{x_j}$ includes all other simplices (as faces: immediate access). Eccentricity is a local index attached at each simplex. Considering a simplex $\overline{x_j}$ of dimension $\dim(\overline{x_j})$ and which higher q-connectivity is of degree $\dim'(\overline{x_j})$, we define the eccentricity of $\overline{x_j}$ by:

$$\eta(\overline{x_j}) = \frac{\dim(\overline{x_j}) - \dim'(\overline{x_j})}{\dim'(\overline{x_j}) + 1} \tag{7}$$

The eccentricity of $\overline{x_j}$ is maximal if it is only connected to the other simplices by a 0-path: its value is then $\eta(\overline{x_j}) = \dim(\overline{x_j})$. $\eta(\overline{x_j}) = 0$ if $\overline{x_j}$ is a sub-simplex (say if there is a $\overline{x_k}$ such that $\overline{x_j} \subset \overline{x_k}$). By convention, we set $\eta(\overline{x_j}) = -1$ if $\overline{x_j}$ is an isolated simplex.

5. Governance ideal types *versus* empirical types

To better understand the specificities of the system we are studying, we propose four models of comparisons corresponding to limiting types of organization of governance, say of ideal types. In the following examples, we assume the same number $N = 8$ or organizations and competences. The global indexes of the corresponding complexes are summarized in **Table** 2. Note that for the no-dependency, full dependency and cyclic ideal types, the incidence matrices are symmetric so that the properties of the simplicial complex K_X and of its conjugate K_Y are the same.

5.1. Ideal type 1: no dependency K^{nodep}

In this model, each of the N organizations has a single competence (works on a single theme) and there is no overlap in the areas of competence of the organizations. This governance structure induces a unitary diagonal $[N \times N]$ square matrix (identity matrix). Each organization is a 0-simplex (a single vertex) with eccentricity -1 since there is no path between organizations (each organization is isolated). The simplicial complex K^{nodep} is of zero dimension $\dim K^{nodep} = 0$. The vector of structure Q also includes only one component equal to the

	dimK	G_{SI}	$\overline{G_{SI}}$	$Q(K)$	$\eta(\overline{x_j})$	$M_{SI}(\overline{x_j})$	$P_q(\overline{x_j})$
$K^{nodep}_{X\ or\ Y}$	0	8	100	[8]	−1	0	0
$K^{fulldep}_{X\ or\ Y}$	7	1	0	[1, 1, 1, 1, 1, 1, 1, 1]	0	1	1
$K^{vertical}_{X}$	7	1	0	[1, 1, 1, 1, 1, 1, 1, 1]	{j = 1}:7 {j≠1}:0	{j = 1}:0.22 {j≠1}:0.11	{j = 1, q = 0}:1 {j ≠1, q = 0}:0.43 {j = 1, q ≥ 1}:0
$K^{vertical}_{Y}$	1	5	57.1	[1, 7]	{j = 1}:0 {j≠1}:1	{j = 1}:0.62 {j≠1}:0.67	{q = 0}:1 {q = 1}:0
$K^{cycle1}_{X\ or\ Y}$	1	5.7	66.7	[1, 8]	1	0.67	{q = 0}:0.56 {q = 1}:0
$K^{cycle2}_{X\ or\ Y}$	2	4.5	50.0	[1, 8]	1/2	0.83	{q = 0}:0.71 {q = 1}:0.56 {q = 2}:0
$K^{cycle3}_{X\ or\ Y}$	3	3.8	40.0	[1, 8]	1/3	0.90	{q = 0}:0.90 {q = 1}:0.71 {q = 2}:0.56 {q = 3}:0

The values of j and q are specified (between braces) only if the value of the index considered is related to them. In the $K^{vertical}_{X}$ complex, j = 1 corresponds to the generalist organization $\overline{x}_{VI}$, which has all the competences.

Table 2. Indices and structure vectors of simplicial complexes corresponding to main ideal type of governance (assuming 8×8 incidence matrices)—see text.

global size index G_{SI} of the complex. This index is equal to the number of independent organizations considered $G_{SI} = N$ (and thus $Q = [N]$) and the maximum dispersion $\overline{G_{SI}} = 100\%$.

5.2. Ideal type 2: full interdependency $K^{fulldep}$

Here, on the contrary, the organizations all work on all the themes and are thus fully interdependent, with no structural leadership. The incidence matrix $[N \times N]$ is full of 1. The dimension of the complex of the organization is determined by the number of themes, $\dim K_X^{fulldep} = (N-1)$. The complex has only one component, all the organizations being connected by an (N–1) path. Each simplex has dimension $(N-1)$ and zero eccentricity: indeed, each simplex coincides with each of the other simplices. The vector of structure $Q\left(K_X^{fulldep}\right)$ is an all-one vector of N components. However, the amalgam of the organizations in a compact structure is expressed by the value of the size index $G_{SI} = 1$ (and dispersion index $\overline{G_{SI}} = 0$), and therefore does not depend on the number N of themes. By symmetry, the conjugate complex $K_Y^{fulldep}$ has similar properties.

5.3. Ideal type 3: vertical integration $K^{vertical}$

For comparison with the other models, in this ideal type, we also consider that the number of organizations is equal to the number of competences N. In the vertical integration model, one of the organizations x_{VI}, integrates all the skills the other organizations having only one of these skills, each time different from the skill of the other organizations. The corresponding incidence matrix is an identity $[N \times N]$ matrix with the addition of the first line (corresponding to the integrative organization) composed of unit elements. The dimension of the simplicial complex $K_X^{vertical}$ of the organizations is given by the dimension of the organization x_{VI} which integrates all the competences ([N – 1]-simplex), say $\dim K_X^{vertical} = (N-1)$. All other organizations are 0-simplices attached to $\overline{x}_{VI}$ by the competence that each one shares with it: there is only one 0-path that binds all organizations. They all have zero eccentricity, being a face (of dimension zero) of the simplex $\overline{x}_{VI}$ with eccentricity $\eta(\overline{x}_{VI}) = \dim(\overline{x}_{VI})$. The structure vector Q_X has N unit components and the global size index is $G_{SI} = 1$ (and dispersion $\overline{G_{SI}} = 0$).

While the diagram associated with $K_X^{vertical}$ consists mainly of the simplex $\overline{x}_{VI}$ of the integrative organization (the other organizations coinciding with its vertices), the diagram of $K_Y^{vertical}$ is a star diagram (a single connected component). The matrix of incidence of the relation $\mathcal{R}^{-1}$ is an identity matrix with the first column filled with 1s. The theme addressed only by x_{VI} is a 0-simplex and the others are all 1-simplices (addressed by x_{VI} and one and only one other organization), thus $\dim K_Y^{vertical} = 1$. Its structure vector has two components $Q_Y = [1, N-1]$ (a 0-simplex and $N-1$ 1-simplices). $K_Y^{vertical}$ has only one connected component (star diagram, no isolated vertex). The eccentricity values do not depend on the size of the network: the eccentricity is zero for the theme taken into account only by the organization x_{VI}, and 1 for the themes covered by two organizations (x_{VI} and one and only one other organization). The global size index depends on the number of organizations with $G_{SI} = 2(N-1)/3$. Finally, it should be noted that in this model, horizontal integration (a competence shared by all organizations, other competences being held by only one organization) is obtained by simply transposing the incidence matrix associated with vertical integration (with simplicial complex $K_X^{vertical}$).

5.4. Ideal type 4: cyclic integration $K^{cycle\theta}$

Let us suppose that we have N themes $\{y_1, y_2, \dots, y_N\}$ and that each organization has competencies on the same number $k < N$ of themes but so that the first organization covers the themes $\{y_1, y_2, \dots, y_k\}$, the second one the themes $\{y_2, y_3, \dots, y_{k+1}\}$ and so on till the last organization with competences on $\{ y_N, y_1, \dots, y_{k-1}\}$. The corresponding complex is formed from N simplices of dimension $(k-1)$ connected by two along a (k − 2)-path forming a cycle. We shall say that this cycle has a *thickness* $\theta = (k-1)$. Two organizations opposite to each other on the cycle have no common focus (theme and competence). However, they are connected by the (k − 1)-path, but separated by a hole. They may be led to dialog, but through other neighboring organizations (with whom they share themes of interest), with some themes being shared between contiguous organizations along this path. As R. Atkin notes, the hole in the middle of the cycle is not just the absence of common competences between opposite organizations on the cycle: it is a real obstacle to cooperation (viewed from the sharing of competences). **Table** 2 presents the values of the indexes for three cyclic models with respective thickness $\theta = 1$, 2 and 3 (assuming again 8×8 incidence matrices). The dimension of the cycle complexes is given by $\dim K_{X\ or\ Y}^{cycle\theta} = \theta$. The θ first values of the structure vector are 1 s, and the last $(\theta+1)^{th}$ component equals N. All simplices have the same eccentricity $\eta = \theta^{-1}$. For a given value of θ, all the simplices have the same meso-index of size M_{SI}. The path index $P_q(\overline{x_j})$ takes quantized values and follows a pattern when changing θ and q as can be seen in **Table** 2.

6. The role of generalist organizations

Consider now the complex K_X^{all} and its conjugate K_Y^{all} representing the organizations involved in Southeast Asia with the distribution of their competences in environment and health as given in **Table** 1. With expertise in each of the areas we are interested in, we consider FAO (Food and Agriculture Organization of the United Nations) and MRC (Mekong River Commission) in this context as *generalist* organizations. The fact that they cover all the competences has several consequences: (a) whatever the dimension threshold considered, the complex has only one connected component (the vector of structure is a all-one vector; it is also the case with K_X^{global} and K_X^{Mekong} for the same reason; **Table** 3); (b) all other organizations are faces, of the FAO-MRC simplex with no dispersion of the organizations $\overline{G_{SI}} = 0$; (c) the eccentricity of all organizations of K_Y^{all} is zero; (d) the dimension values of the simplices and their meso-index of size M_{SI} are congruent (provide the same information) as seen in **Figure 2**. Overall, the K_X^{all} simplicial complex is very similar to the ideal type of vertical integration $K_X^{vertical}$.

No competence to solicit all organizations, the structure of the conjugate complex K_Y^{all} is less homogeneous. Two factors contribute to a high value of the mesoindex of size: a high number of vertices connected by q-geodesics of maximum length and in addition that this degree q is low—see Eq. (5). This is the case of simplices $\overline{HH}$ (human health), $\overline{RA}$ (risk assessment) and $\overline{BD}$ (biodiversity) (**Figure 3**). The eccentricity varies according to the competence considered. In this context of governance, the skill regarding biodiversity is more eccentric, less integrated to the set of other competences. Indeed $\overline{RA}$ and $\overline{BD}$ are of the same dimension (18) but the degrees of q-connectivity are q = 14 for $\overline{RA}$ and q = 10 for $\overline{BD}$; $\overline{HH}$ is of dimension 19 and of higher q-connectivity q = 14.

The structure vector Q_Y^{all} also contains very useful information. It indicates the number of co-existing cliques when only the simplices of a dimension greater or equal to a threshold dimension are maintained. In the case of K_Y^{all}, the cliques are represented in the Q-analysis tree in **Figure 4**. The lower dimensional simplices (disappearing first from the tree of cliques) are land use ($\overline{LU}$), then animal health ($\overline{AH}$) and ecosystem health ($\overline{EH}$). $\overline{AH}$ and $\overline{EH}$ are also the first simplices to dissociate from the main clique. These properties show that animal health and

$[M, N]$	Complex	$\dim K_X$	$\overline{G_{SI}}$	$Q(K)$	$Q(K)$	$\overline{G_{SI}}$	$\dim K_Y$	Complex
[34, 9]	K_X^{all}	8	0	[1, 1, 1, 1, 1, 1, 1, 1, 1]	[1, 1, 1, 1, 1, 1, 2, 3, 2, 1, 2, 4, 3, 3, 3, 3, 3, 3, 3, 1]	18.0	19	K_Y^{all}
[5, 9]	K_X^{global}	8	0	[1, 1, 1, 1, 1, 1, 1, 1, 1]	[1, 2, 1, 1]	2.5	3	K_Y^{global}
[7, 9]	K_X^{ASEAN}	6	0.6	[2, 1, 1, 1, 1, 1, 1]	[2, 1, 1, 3]	11.2	3	K_Y^{ASEAN}
[5, 9]	K_X^{SEAMEO}	4	28.3	[1, 2, 3, 2, 2]	[1, 3, 1, 1]	5.0	3	K_Y^{SEAMEO}
[7, 9]	K_X^{Mekong}	8	0	[1, 1, 1, 1, 1, 1, 1, 1, 1]	[1, 1, 1, 3, 2]	10.8	4	K_Y^{Mekong}
[10, 9]	K_X^{AsiPac}	5	7.9	[1, 1, 2, 4, 1, 1]	[1, 1, 3, 4, 2, 1]	13.7	5	K_Y^{AsiPac}

Table 3. Global indices and structure vectors of various complexes corresponding to empirical types of health-environment governance.

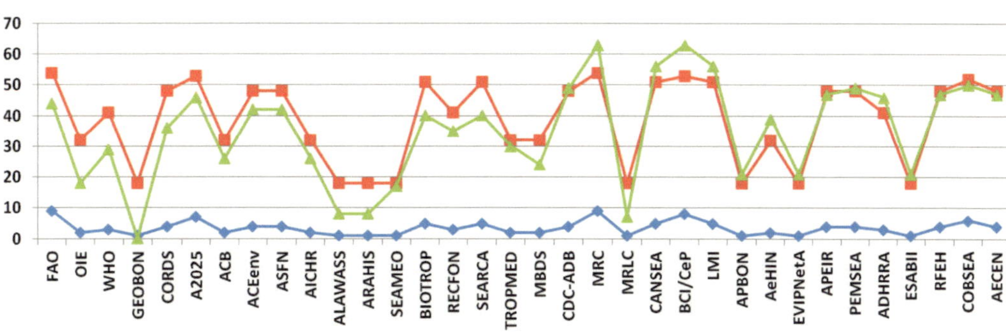

Figure 2. Values of the dimension + 1 (diamonds) and meso-index of size M_{SI} (squares) for the simplices (organizations) of K_X^{all}. The values of the meso-index of size obtained by considering each organization group separately (global, ASEAN, SEAMEO, organizations of the Mekong Basin, Asia-Pacific organizations—see **Table** 1) are also represented (triangles).

ecosystem health skills are the least well integrated in this context of regional governance, while their integration with human health is central to the One Health approach. Similarly, land use skills are very important—especially if they are linked to epidemiological competences—the life cycle of several vectors and pathogens being influenced by land use and land cover changes [36]. Finally, the clique of competences that we can classify under the label *environmental changes* (climate change, biodiversity, water resources, food security) also dissociates quite quickly (in dimension 10), revealing an institutional gap between these competencies (in this context again).

The Q-analysis can be done by considering in turn each subgroup of organizations—global organizations, ASEAN, SEAMEO, Mekong Basin and Asia-Pacific organizations (see **Table** 1). The trees showing the fragmentation of the competence cliques with the increase of the

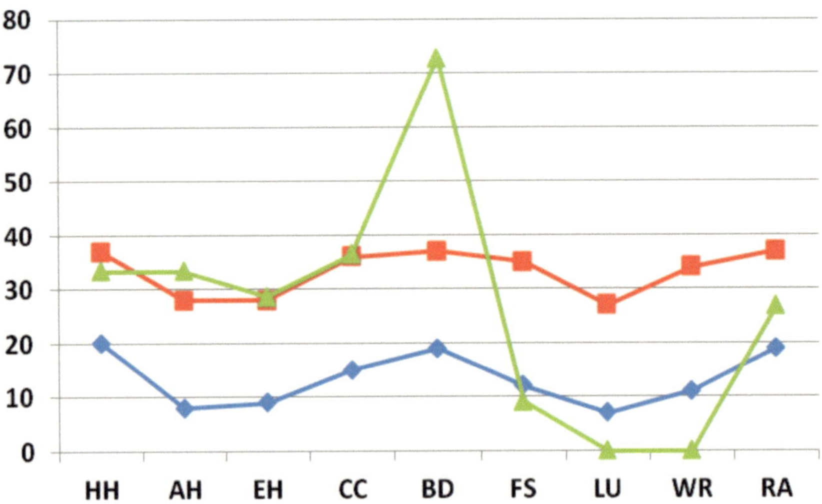

Figure 3. Number of vertices (diamonds), meso-size index $M_{SI} \times 100$ (squares) and eccentricity $\eta \times 100$ (triangles) of the complex of competences K_Y^{all}.

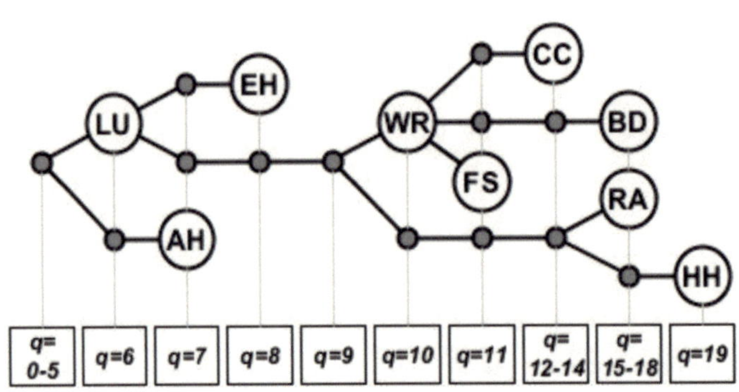

Figure 4. Q-analysis tree of K_Y^{all}: clique of competences (with labels given in **Table 1**) as a function of the threshold dimension q, with structure vector $Q(K_Y^{all}) = [1,1,1,1,1,1,2,3,2,1,2,4,3,3,3,3,3,3,3,1]$. The threshold dimension q is indicated in the bottom boxes.

threshold dimension are very different from each other (**Figure 5**) and do not make it possible to infer *a priori* that which results from their association in **Figure 4**. At the beginning, each group presents all the competences distributed among its member organizations (except ASEAN without competence in environmental health). But according to the organizations involved, each competence is more or less shared in the group. The main ones (at the top of the trees) will tend to promote the associated theme as one that federates the activities of the organization group: risk assessment for global organizations, human health for Asia-Pacific and SEAMEO, the importance of climate change for ASEAN, etc. Groups with the most member organizations tend to have higher competence trees (5 for Asia-Pacific). It is also observed that although the Global and Mekong groups have each a generalist organization (FAO and MRC respectively), the competence cliques are not comparable.

Of course, the association of all these groups produces a higher clique tree (q = 19, **Figure 4**), with a more robust network of competences with respect to a change of skill, or even the discontinuance of an organization. The integration of groups in the regional governance system has differentiated effects for each organization. In **Figure 2**, it is observed, for example, that the meso-index of size M_{SI} decreases for the MRC generalist organization, whereas it increases for the FAO. GEOBON's relative size decreases with this integration, whereas that of APBON (both dedicated to the management of biodiversity observations) remains unchanged. The competence portfolio (and hence the number of vertices) remains unchangesd by the integration of organizations, thus any change in the size meso-index reflects the

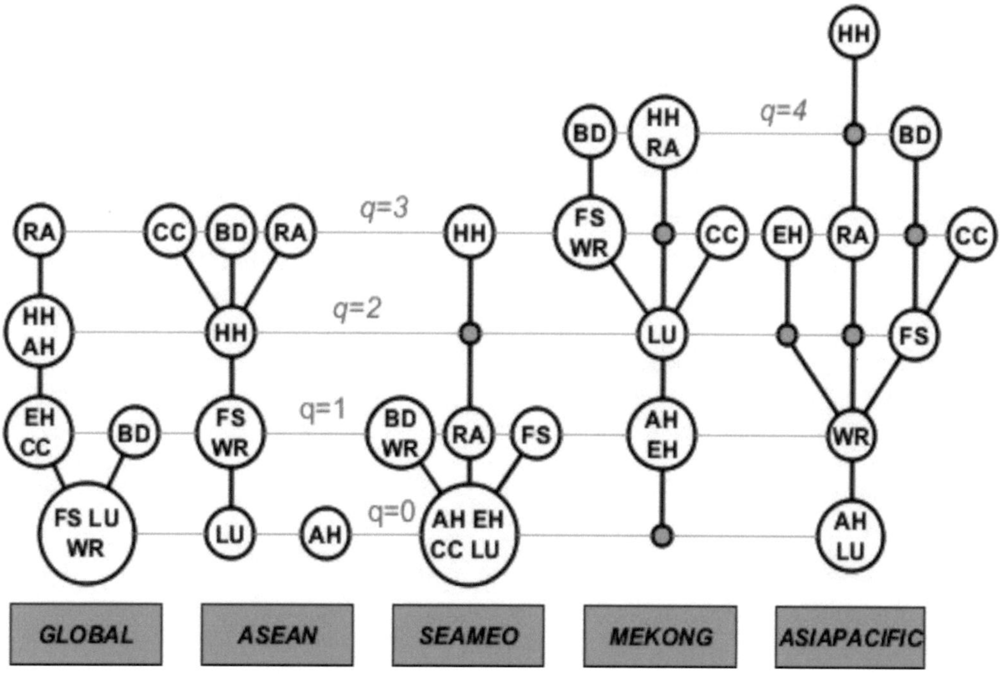

Figure 5. Q-analysis tree of cliques of competences considered separately for each group of organizations (labels in the rectangular boxes at the bottom). The threshold dimension q is indicated for each level.

modifications of the q-paths, the low-dimensional ones being weighted more in this index (cf. Eq. (5)).

Without going into details, the tree of the cliques of organizations obtained according to the threshold dimension is less interesting in this context than that of competences presented in **Figure 5**. Indeed, the tree associated with each organization group resembles more or less that corresponding to another ideal type. This one, which we call *pyramidal* ideal type (inverted), is composed of $n = 1 \ldots N$ organizations, the nth having $(N - n + 1)$ competencies. At the top there is a generalist organization and at the bottom an organization presenting only one competence. In this situation, the tree of the cliques has only one trunk that loses an organization with each unit increment of the threshold dimension. The path indexes (Eq. (6)) of almost all organizations change with their integration in the larger "all" governance system as can be seen for the FAO and COBSEA (Coordinating Body on the Seas of East Asia) organizations and for the strategic policy program ASEAN2025 [35] in **Figure 6**.

The change in the P_q path indexes expresses the fact that in general the integration of an organization in a large governance systems multiplies the q-paths and the opportunities to find some potential partners with similar competences and interest in common themes. Of course, a generalist organization like FAO takes maximum advantage of such integration. But it is also interesting to see that a political strategy as expressed by the ASEAN countries in their ASEAN 2025 policy [35] offers new perspectives and new connectivity, when considered in a broader governance context. Similarly, an organization like COBSEA, focused on issues related

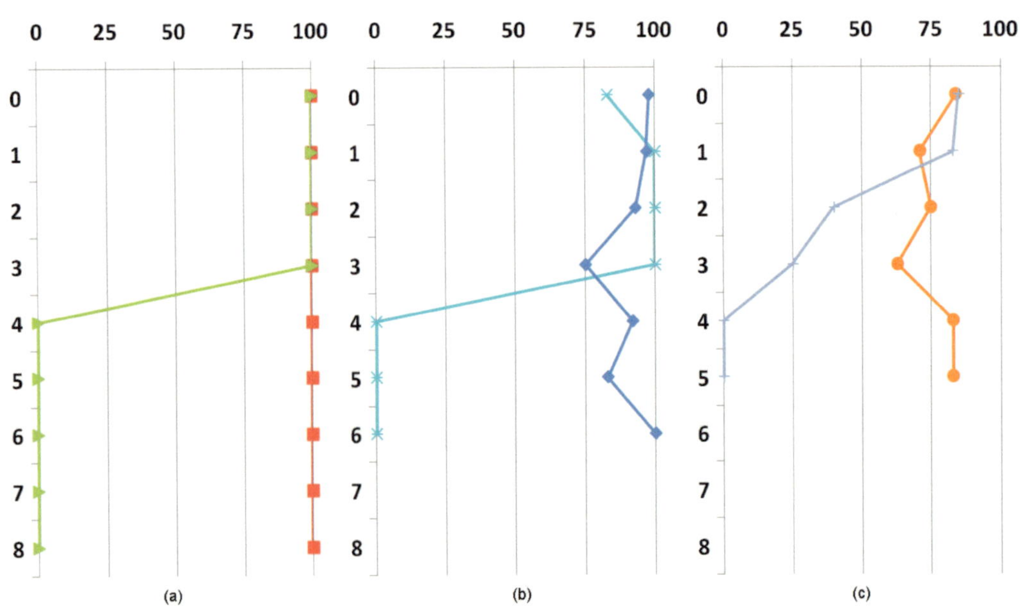

Figure 6. Path indexes (×100; x-axis) of FAO, ASEAN2025 policy text and COBSEA as a function of the threshold dimension (y-axis). Each entity is considered both in the K_X^{all} simplicial complex and in the complex corresponding to their organizations group (see text and **Table 1**). (A) FAO [squares: in K_X^{all}; triangles: in K_X^{global}]; (B) ASEAN2025 [diamonds: in K_X^{all}; crosses: in K_X^{asean}] and (C) COBSEA [dots: in K_X^{all}; plus: in K_X^{asipac}].

http://dx.doi.org/10.5772/intechopen.71774

to the management of regional seas and marine resources and environments, is to be reconsidered in the larger governance system, as it is true that the relative position that each occupies depends closely on the context taken into account. All the information produced by Q-analysis is not exploited here, but position and importance of attributes of each entity—actors (e.g. organizations) and framework for action (policy strategy, legal instruments, etc.)—can be analyzed according to different governance contexts where it integrates or wishes to integrate.

7. Discussion

The mathematical concepts used in this article remain elementary, but it is important to note that the two conjugate complexes associated with the same relation, even though they have generally very different combinatorial appearances, share strong topological properties. From the mathematical point of view, this is reflected in the identity of their homology groups and their homotopy groups [37, 38]. This goes well beyond the elementary considerations to which we limit ourselves here in our modeling but the identity of these topological characteristics reinforces the importance of the principle of conjugacy between the two simplicial complexes naturally associated with a given relation.

For governance studies, the interest of such an approach is that it allows understanding very different contexts of governance by describing the actors and organizations already into action and the way they connect to each other. Ultimately, it also makes possible to delineate the institutions and issues at stake and to highlight the different levels of decision-making and thus of regulations involved. It can apply in various settings. For instance, one of the issues underlined by Osofsky [1] in the case of the environmental disaster resulting from the BP Deepwater Horizon oil spill is the need for integration across scales. The spill stretched over the shoreline of five states of the United States, and due to the multiplicity of decision levels (local and federal governments) and the variety of institutions involved (such as the Department of Agriculture, Department of Defense, Department of Energy, Department of Homeland Security, Department of Justice, Department of the Interior, Department of Labor, Environmental Public Agency, Health and Human Services or National Aeronautics and Space Administration …), one of the legal difficulties was to disentangle the overlaps of regulations or on the contrary the gaps resulting from the legal fragmentation.

The approach can thus be used in this kind of context or either to determine in a specific area, like the Southeast Asian region, how the health and environmental governance works to identify the missing linkages or the possibilities for synergies. It is a flexible approach and the results and their interpretations are depending on the context chosen as well as on the organizations, networks and themes considered in the research scope. This flexibility can be seen not only as a limit of the approach but also as an advantage as it allows to change the analysis framework: in a first phase, we could choose to consider a specific type of organizations (in a predetermined typology) and thus extend the research to other types of organizations. It is particularly relevant when it comes to describing and interpreting multidimensional and multilevel interactions.

The modeling approach is also very useful when governance systems are composed of hundreds of organizations and tens of attributes or when the ambition is to simulate the impact of changes of governance structure through scenarios. System wide indexes (global indexes), local indexes attached to each organization or attribute and meso-indexes assessing how they are inserted are exploited not only to construct global governance diagnoses but also to follow each entity in the evolving governance architecture.

On a semantic point of view, the use of the term *model* itself in the legal or political arena is different than in mathematics, physics or computer sciences. This can have methodological repercussions, as the term "model" can be used to define a descriptive approach closer to an enumeration of facts than to a systemic approach. Indeed, when speaking about models of governance, legal scholars usually refers to an analytical or normative framework rather than to a model integrating interactions and showing a dynamic expressed through mathematical properties translating types of behaviors or linkages. Nevertheless, this type of formal model opens the perspective of many analyzes of real systems of governance seen from new and diversified angles.

8. Conclusion

We have enriched our analytical tools with another approach to modeling systemic governance based on Q-analysis and using the simplicial complexes as a mathematical object (type of hypergraph or hyper-network). The model allows taking into account a variety of entities as elements of governance, say organizations, networks (of networks) of organizations, technical platforms, but also legal instruments (e.g. norms, agreements) and public policies. Since these entities can be characterized in different ways, modeling leads us to consider governance under as many different angles as there are types of attributes associated with entities.

The simplicial complexes introduce formal concepts of dimension and conjugacy between the hyper-network of entities (e.g. organizations) and the hyper-network formed by a choice of attributes, the two simplicial complexes being bound by topological properties. Several indicators are evaluated to characterize the global (overall), mesoscale and local (at the scale of each organization or attribute) properties of each of the two conjugated complexes associated with a given context of governance. Moreover, these indicators also make it possible to compare distinct systems of governance. Thus, we have also established the indices associated with several ideal type of governance that stylizes limit situations between organizations (or other entities): complete independence, full interdependence, vertical integration and horizontal integration and cyclic governance. The flexibility of the analytical tool makes it suitable for exploring a wide variety of governance systems, the case discussed in more detail here considering groups of organizations involved in Southeast Asia on health-environment issues.

Acknowledgements

This study contributes to the International Multidisciplinary Thematic Network *Biodiversity, Health and Societies in Southeast Asia*, Thailand supported by the Ecology and Environment Institute of the National Centre for Scientific Research (InEE CNRS, France). It is supported by the FutureHealthSEA Project "*Predictive scenarios of health in Southeast Asia: linking land use and climate changes to infectious diseases*" (funded by ANR 2017) and by the GEMA project "*Gouvernance Environnementale: Modélisation et Analyse*" (funded by CNRS Défi interdisciplinaire InFIniti).

Author details

Pierre Mazzega[1]*, Claire Lajaunie[2] and Etienne Fieux[3]

*Address all correspondence to: pierre.mazzegaciamp@get.omp.eu

1 Geosciences Environment Toulouse UMR5563, CNRS, University of Toulouse, Toulouse, France

2 INSERM DICE-Ceric UMR7318—International, Comparative and European Law, Aix-Marseille University, Aix en Provence, France

3 Institute of Mathematics of Toulouse UMR 5219, University of Toulouse, Toulouse, France

References

[1] Osofsky HM. Multidimensional governance and the BP Deepwater horizon oil spill. Florida Law Review. 2011;**63**(5):1077-1137. Available from: http://scholarship.law.ufl.edu/flr/vol63/iss5/2 Accessed: 2017-09-24

[2] Peel J, Osofsky HM. Climate Change Litigation: Regulatory Pathways to Cleaner Energy. Cambridge: Cambridge University Press; 2015. 352 p. ISBN: 9781107036062

[3] Müller-Mall S. Legal Spaces. Towards a Topological Thinking of Law. Heidelberg: Springer; 2013. x+132 p. DOI: 10.1007/978-3-642-36730-4

[4] Badii R, Politi A. Complexity. Hierarchical Structures and Scaling in Physics. Nonlinear Sci. Ser. 6. Cambridge: Cambridge University Press. 2013. xiii + 317 p. ISBN: 0-521-66385-7

[5] Kantz H, Schreiber Th. Nonlinear Time Series Analysis. Nonlinear Sci. Ser. 7. Cambridge: Cambridge University Press. 1997. xi + 304 p. ISBN: 0-521-55144-7

[6] Levi-Faur D, editor. The Oxford Handbook of Governance. Oxford: Oxford University Press; 2012. xxii + 803 p. ISBN: 978-0-19-967706-1

[7] Box-Steffensmeier JM, Brady HE, Collier D, editors. The Oxford Handbook of Political Methodology. Oxford: Oxford University Press; 2008. xiii + 880 p. ISBN: 978-0-19-958556-4

[8] Reus-Smit C, Snidal D, editors. The Oxford Handbook of International Relations. Oxford: Oxford University Press; 2008. xiii + 772 p. ISBN: 978-0-19-958558-8

[9] Munkres JR. Elements of Algebraic Topology. Menlo Park: Addison-Wesley Publ. Company; 1984. ix + 454 p. ISBN: 0-201-04586-9

[10] Kozlov D. Combinatorial Algebraic Topology. Algorithms and Computation in Mathematics, 21. Berlin: Springer; 2008. xx + 389 p. ISBN: 978-3-540-71961-8

[11] Lajaunie C, Mazzega P. Organization networks as information integration system—Case study on environment and health in Southeast Asia. Advances in Computer Science: An International Journal. 2016;**2**(20):28-39. Available from: http://www.acsij.org/acsij/article/view/461 Accessed: 2017-09-24

[12] Mazzega P, Lajaunie C. Modeling organization networks collaborating on health and environment within ASEAN. In: Martinez RS, editor. Complex Systems: Theory and Applications. Hauppauge: NOVA Science Publ.; 2017. p. 117-148. ISBN: 978-1-53610-871-2

[13] Atkin RH. From cohomology in physics to q-connectivity in social sciences. International Journal of Man-Machine Studies. 1972;**4**:139-167

[14] Atkin RH. Q-analysis—A hard language for the soft sciences. Futures. 1978;**10**(6):492-499

[15] Atkin RH. Combinatorial Connectivities in Social Systems. Basel: Springer; 1977. ii + 241 p. ISBN: 978-3-7643-0912-1

[16] Gould P. Q-analysis, or a language of structure: An introduction for social scientists, geographers and planners. International Journal of Man-Machine Studies. 1980;**13**:169-199

[17] Duckstein L, Nobe SA. Q-analysis for modeling and decision making. European Journal of Operational Res. 1977;**103**:411-425

[18] Jiang B, Omer I. Spatial topology and its structural analysis based on the concept of simplicial complex. Transactions in GIS. 2007;**11**(6):943-960

[19] Johnson J. Hyper-Networks in the Science of Complex Systems. London: Imperial College Press—World Scientific Publishing Co. Pte. Ltd.; 2013. xviii + 330 pp. ISBN: 978-1-86094-972-2

[20] Maletic S, Rajkovic M. Consensus formation on a simplicial complex of opinions. Physica A: Statistical Mechanics and its Applications. 2014;**397**:111-120. DOI: https://doi.org/10.1016/j.physa.2013.12.001

[21] Kee KF, Sparks L, Struppa DC, Mannucci M, Daminano A. Information diffusion, facebook clusters, and the simplicial model of social aggregation: A computational simulation of simplicial diffusers for community health interventions. Health Communication. 2016;**31**(4):385-399. DOI: 10.1080/10410236.2014.960061

[22] Estrada E, Ross G. Centralities in Simplicial Complexes. 2017. ArXiv: 1703.03641

[23] Carlsson G. Topology and data. Bulletin of the American Mathematical Society. 2009; **46**(2):255-308. DOI: https://doi.org/10.1090/S0273-0979-09-01249-X

[24] Barcelo H, Laubenbacher R. Perspectives on A-homotopy theory and its applications. Discrete Mathematics. 2005;**298**(1–3):39-61 https://doi.org/10.1016/j.disc.2004.03.016

[25] Morand S, Lajaunie C. Biodiversity and Health: Linking Life, Ecosystems and Societies. Elsevier ISTE Press; ISBN-13: 978-1785481154

[26] AVMA. One Health—A New Professional Imperative. American Veterinary Medical Assoc; 2008. 71 p Accessed: 2017-09-24. Available from: https://www.avma.org/KB/Resources/Reports/Documents/onehealth_final.pdf

[27] Bousfield B, Brown R. One world one health. veterinary bulletin—Agriculture, fisheries and conservation department. Newsletter. 2011;**1**(7):1-12. Available from: https://www.afcd.gov.hk/tc_chi/quarantine/qua_vb/files/OWOH2.pdf [Accessed: 24-09-2017]

[28] Field CB, Barros V, Dokken DJ, et al. Climate change 2014: Impacts, adaptation, and vulnerability. Volume I: Global and Sectoral Aspects. Contribution of Working Group II to the Fifth Assessment Report of the Intergovernmental Panel on Climate Change. 2014. Available from: http://www.ipcc.ch/report/ar5/wg2/ [Accessed: 2017-09-24]

[29] Watts N, Adger WN, Agnolucci P, et al. Health and climate change: Policy responses to protect public health. The lancet commissions. The Lancet. 2015;**386**:1861-1914 http://dx.doi.org/10.1016/S0140-6736(15)60854-6

[30] Morand S, Jittapalapong S, Supputamongkol Y, Abdullah MT, Huan T. Infectious diseases and their outbreaks in Asia-Pacific: Biodiversity and its regulation loss matter. PLoS One. 2014;**9**(2):e90032 https://doi.org/10.1371/journal.pone.0090032

[31] McIntyre KM, Setzkorn C, PhJ H, Morand S, Moser AP, Baylis M. Systematic assessments of the climate sensitivity of important human and domestic animals pathogens in Europe. Scientific Reports. 2017:1-10. DOI: 10.1038/s41598-017-06948-9

[32] Jones KE, Patel NG, Levy MA, Storeygard A, Balk D, Gittleman JL, Daszak P. Global trends in emerging infectious diseases. Nature. 2008;**451**:990-994. DOI: 10.1038/nature06536

[33] Lajaunie C, Mazzega P, Boulet R. Health in biodiversity-related conventions: Analysis of a multiplex terminological network (1973–2016). In: Chen S-H, editor. Big Data in Computational Social Science and Humanities. Heidelberg: Springer; 2017 (in press)

[34] Lajaunie C, Mazzega P. One health and biodiversity conventions. The emergence of health issues in biodiversity conventions. IUCN Academy of Environmental Law eJournal. 2016;**7**:105-121. Available from: www.iucnael.org/en/documents/1299-one-health-and-biodiversity-conventions Accessed: 2017–09-24

[35] ASEAN. ASEAN2025: Forging Ahead Together. The ASEAN Secretariat: Jakarta, Indonesia; 2015. 136 p Available from: www.asean.org/storage/2015/12/ASEAN-2025-Forging-Ahead-Together-final.pdf Accessed: 2017-09-24978-602-0980-45-4

[36] Patz JA, Daszak P, Tabor GM, Aguirre AA, Pearl M, Epstein J, Wolfe ND, Kilpatrick AM, Foufopoulos J, Molyneux D, Bradley DJ. Unhealthy landscapes: Policy recommendations

on land use change and infectious disease emergence. Environmental Health Perspectives. 2004;**112**:1092-1098. DOI: 10.1289/ehp.6877

[37] Dowker CH. Groups of relations. Annals of Mathematics. 1952;**56**(1):84-95

[38] Minian EG. The geometry of relations. Order. 2010;**277**:213-224. DOI: 10.1007/s11083-010-9146-4

Section 2

Graph Theory in Depth

Chapter 5

Modeling Rooted in-Trees by Finite *p*-Groups

Daniel C. Mayer

Additional information is available at the end of the chapter

http://dx.doi.org/10.5772/intechopen.68703

Abstract

Graph theoretic foundations for a kind of infinite rooted in-trees $\mathcal{T}(R) = (V, E)$ with root R, weighted vertices $v \in V$, and weighted directed edges $e \in E \subset V \times V$ are described. Vertex degrees $\deg(v)$ are always finite but the trees contain infinite paths $(v_i)_{i \geq 0}$. A concrete group theoretic model of the rooted in-trees $\mathcal{T}(R)$ is introduced by representing vertices by isomorphism classes of finite p-groups G, for a fixed prime p, and directed edges by epimorphisms $\pi: G \to \pi G$ of finite p-groups with characteristic kernels $\ker(\pi)$. The weight of a vertex G is realized by its nuclear rank $n(G)$ and the weight of a directed edge π is realized by its step size $s(\pi) = \log_p(\#\ker(\pi))$. These invariants are essential for understanding the phenomenon of *multifurcation*. Pattern recognition methods are used for finding finite subgraphs which repeat indefinitely. Several periodicities admit the reduction of the complete infinite graph to finite patterns. The proof is based on infinite limit groups and successive group extensions. It is underpinned by several explicit algorithms. As a final application, it is shown that *fork topologies*, arising from repeated multifurcations, provide a convenient description of complex navigation paths through the trees, which are of the greatest importance for recent progress in determining p-class field towers of algebraic number fields.

Keywords: rooted directed in-trees, descendant trees, infinite paths, vertex distance, weighted edges, pattern recognition methods, pattern classification, independent component analysis, graph dissection, finite p-groups, projective limits, periodicity, group extensions, nuclear rank, multifurcation, presentations, commutators, central series

1. Introduction

In Section 2, we describe the abstract graph theoretic foundations for a kind of infinite rooted in-trees $\mathcal{T}(R) = (V, E)$ with root R, weighted vertices $v \in V$, and weighted directed edges

$e \in E \subset V \times V$, which are suited perfectly for describing the crucial phenomenon of *multifurcation* in Section 2.3. The vertex degrees $\deg(v)$ are always finite, but the trees contain infinite paths $(v_i)_{i \geq 0}$. In Section 3, we introduce a group theoretic model of the rooted in-trees $\mathcal{T}(R)$. Vertices are represented by isomorphism classes of finite p-groups G, for a fixed prime number p. Directed edges are represented by epimorphisms $\pi : G \to \pi G$ of finite p-groups with characteristic kernels $\ker(\pi)$. The weight of a vertex G is realized by its *nuclear rank* $n(G)$, and the weight of a directed edge π is realized by its *step size* $s(\pi) = \log_p(\#\ker(\pi))$. Since the structure of our rooted in-trees is rather complex, we use *pattern recognition* methods in Section 3.1 for finding finite subgraphs which repeat indefinitely as branches of coclass subtrees, thus giving rise to a *first periodicity*. Additionally, we employ *independent component analysis* for obtaining a graph dissection into pruned subtrees, either by Galois action in Section 3.2.1 or by Artin transfers in Section 3.2.2. A *second periodicity* of pruned coclass subtrees eventually admits the reduction of the complete infinite graph $\mathcal{T}(R)$ to finite patterns. Evidence of these newly discovered *periodic bifurcations* is provided by a mixture of bottom up techniques, using successive extensions by means of p-covering groups in the *p-group generation algorithm* and top down techniques using *infinite limits* and their *finite quotients* in Sections 3.4 and 3.5. As a coronation of this chapter, we show in Sections 3.6 and 4 that *fork topologies* provide a convenient description of very complex navigation paths through the trees, arising from repeated multifurcations, which are of the greatest importance for recent progress in determining *p-class field towers* $F_p^{(\infty)}$ of algebraic number fields F.

2. Underlying abstract graph theory

Let $\mathcal{G} = (V, E)$ be a *graph* with set of *vertices* V and set of *edges* E. We expressly admit infinite sets V and E, but we assume that the in and out degree of each vertex is finite.

2.1. Directed edges and paths

In this chapter, we shall be concerned with *directed graphs* (digraphs) whose edges are rather ordered pairs $(v_1, v_2) \in V \times V$ than only subsets $\{v_1, v_2\} \subset V$ with two elements. Such a *directed edge* $e = (v_1, v_2)$ is also denoted by an arrow $v_1 \to v_2$ with starting vertex v_1 and ending vertex v_2. Thus, we have $E \subset V \times V$. Now, infinitude comes in.

Definition 2.1. (Finite and infinite paths.)

A *finite path* of *length* $\ell \geq 0$ in $\mathcal{G}$ is a finite sequence $(v_i)_{0 \leq i \leq \ell}$ of vertices $v_i \in V$ such that $(v_i, v_{i+1}) \in E$ for $0 \leq i \leq \ell - 1$. We call v_0, respectively, v_ℓ, the starting vertex, respectively, ending vertex, of the path. The degenerate case of a single vertex (v_0) is called a *point path* of length $\ell = 0$.

An *infinite path* in $\mathcal{G}$ is an infinite sequence $(v_i)_{i \geq 0}$ of vertices $v_i \in V$ such that $(v_{i+1}, v_i) \in E$ for all $i \geq 0$. In this case, v_0 is the ending vertex of the path, and there is no starting vertex.

http://dx.doi.org/10.5772/intechopen.68703

2.2. Rooted in-trees with parent operator

Our attention will even be restricted to *rooted in-trees* $\mathcal{G} = \mathcal{T}(R)$, that is, connected digraphs without cycles such that the *root vertex* R has out-degree 0, whereas any other vertex $v \in V \setminus \{R\}$ has out-degree 1. A vertex with in-degree at least 1 is called *capable*, whereas a vertex with in-degree 0 is called a *leaf*. For a rooted in-tree, we can define the parent operator as follows.

Definition 2.2. Let $\mathcal{T}(R) = (V, E)$ be a rooted in-tree. Then, the mapping $\pi : V \setminus \{R\} \to V$, $v \mapsto \pi v$, where $(v, \pi v) \in E$ is the unique edge with starting vertex v, is called the *parent operator* of $\mathcal{T}(R)$. For each vertex $v \in V$, there exists a unique finite *root path* from v to the root R,

$$v = \pi^0 v \to \pi^1 v \to \pi^2 v \to \ldots \to \pi^{\ell-1} v \to \pi^{\ell} v = R,$$

expressed by iterated applications of the parent operator and with some *length* $\ell \geq 0$. Each vertex in the root path of v is called an *ancestor* of v.

The *descendant tree* $\mathcal{T}(a) = (V(a), E(a))$ of a vertex $a \in V$ is the subtree of $\mathcal{T}(R) = (V, E)$ consisting of vertices v with ancestor a, that is $v \in V(a) := \{u \in V \mid (\exists j \geq 0)\ \pi^j u = a\}$, and edges $e \in E(a) := E \bigcap (V(a) \times V(a))$.

A vertex $u \in V$ is called an *immediate descendant* (or *child*) of a vertex $a \in V$, if there exists a directed edge $(u, a) \in E$. In this case, $a = \pi u$ is necessarily the *parent* of u.

We can define a *partial order* on the vertices $u, a \in V$ of the tree $\mathcal{T}(R)$ by putting $u \geq a$ if $u \in \mathcal{T}(a)$, that is, if u is descendant of a and a is ancestor of u. The root R is the minimum.

The root R is always a common ancestor of two vertices $u, v \in V$. By the *fork* of u and v, we understand their biggest common ancestor, denoted by $\mathrm{Fork}(u, v)$, which admits a measure.

Definition 2.3. (Vertex distance.) The sum $\ell_u + \ell_v$ of the path lengths from two vertices $u, v \in V$ to their fork is called the *distance* $d(u, v)$ of the vertices.

2.3. Mainlines and multifurcation

We shall also need *weight functions* with nonnegative integer values for vertices $w_V : V \to \mathbb{N}_0$, and with positive integer values for edges $w_E : E \to \mathbb{N}$. In particular, the sets of vertices and edges have disjoint partitions

$$\begin{aligned} V &= \dot{\bigcup}_{n \geq 0} V_n \text{ with } V_n := \{v \in V \mid w_V(v) = n\} \text{ for } n \geq 0, \\ E &= \dot{\bigcup}_{s \geq 1} E_s \text{ with } E_s := \{e \in E \mid w_E(e) = s\} \text{ for } s \geq 1, \end{aligned} \tag{1}$$

such that V_0 is precisely the *set of leaves* of the tree $\mathcal{T}(R)$. Thus, there arise weighted measures.

Definition 2.4. (Path weight and weighted distance.)

By the *path weight* of a finite path $(v_i)_{0 \le i \le \ell}$ with length $\ell \ge 0$ in $\mathcal{T}(R)$ such that $(v_i, v_{i+1}) \in E_{s_i}$ for $0 \le i \le \ell - 1$, we understand the sum $\sum_{i=0}^{\ell-1} s_i$. The sum $w_u + w_v$ of the path weights from two vertices $u, v \in V$ to their fork is called the *weighted distance* $w(u, v)$ of the vertices.

In Definitions 2.5 and 2.6, some concepts are introduced using the minimal possible weight.

Definition 2.5. (Mainlines and minimal trees.) An infinite path $(v_i)_{i \ge 0}$ in $\mathcal{T}(R)$ with edges of weight 1, that is, such that $(v_{i+1}, v_i) \in E_1$ for all $i \ge 0$, is called a *mainline* in $\mathcal{T}(R)$.

The *minimal tree* $\mathcal{T}_1(a) = (V_1(a), E_1(a))$ of a vertex $a \in V$ is the subtree of the descendant tree $\mathcal{T}(a) = (V(a), E(a))$ consisting of vertices v, whose root path in $\mathcal{T}(a)$ possesses edges e of weight 1 only, that is $v \in V_1(a) := \{u \in V(a) | (\forall\, 0 \le j < \ell)\ (\pi^j u, \pi^{j+1} u) \in E_1\}$, and edges $e \in E_1(a) := E(a) \bigcap (V_1(a) \times V_1(a))$.

Definition 2.6. (Branches.) Let $(v_i)_{i \ge 0}$ be a mainline in $\mathcal{T}(R)$. For $i \ge 0$, the difference set $\mathcal{B}(v_i) := \mathcal{T}_1(v_i) \backslash \mathcal{T}_1(v_{i+1})$ of minimal trees is called the *branch* with root v_i of the minimal tree $\mathcal{T}_1(v_0)$. The branches give rise to a disjoint partition $\mathcal{T}_1(v_0) = \dot{\bigcup}_{i \ge 0} \mathcal{B}(v_i)$.

Finally, we complete our abstract graph theoretic language by considering arbitrary weights.

Definition 2.7. (Multifurcation.)

Let $n \ge 2$ be a positive integer. A vertex $a \in V_n$ has an n-fold *multifurcation* if its in-degree is an n-fold sum $N_1 + N_2 + \ldots + N_n$ due to $N_s \ge 1$ incoming edges of weight s, for each $1 \le s \le n$. That is, we define counters N_s of all incoming edges of weight s, and additionally, we have counters C_s of all incoming edges of weight s with capable starting vertex

$$\begin{aligned} N_s &:= N_s(a) := \#\{e \in E_s | e = (u, a) \text{ for some } u \in V\}, \\ C_s &:= C_s(a) := \#\{e \in E_s | e = (u, a) \text{ for some } u \in V \text{ with } w_V(u) \ge 1\}. \end{aligned} \quad (2)$$

We also define an ordering and a notation [1] for immediate descendants of a by writing $a - \#s; i$ for the ith immediate descendant with edge of weight s, where $1 \le s \le n$ and $1 \le i \le N_s$.

3. Concrete model in *p*-group theory

Now, we introduce a group theoretic model of the rooted in-trees $\mathcal{T}(R) = (V, E)$ in Section 2. Vertices $v \in V$ are represented by isomorphism classes of finite p-groups G, for a fixed prime number p. Directed edges $e \in E$ are represented by epimorphisms $\pi : G \to \pi G$ of finite p-groups with characteristic kernels $\ker(\pi) = \gamma_c G$, where $c := \mathrm{cl}(G)$ denotes the nilpotency class of G and $(\gamma_i G)_{i \ge 1}$ is the lower central series of G.

We emphasize that the symbol π is used now intentionally for two distinct mappings, the abstract parent operator $\pi : V \backslash \{R\} \to V$, $v \mapsto \pi v$, in Definition 2.2, and the concrete natural projection onto the quotient $\pi : G \to \pi G \simeq G/\gamma_c G$, $g \mapsto \pi(g) = g \cdot \gamma_c G$, for each individual vertex $G = v \in V \backslash \{R\}$, which should precisely be denoted by $\pi = \pi_G$, but we omit the subscript, since there is no danger of misinterpretation. In both views, πG is the parent of G.

The weight of a vertex G is realized by its *nuclear rank* $n(G)$ ([2], section 14, eqn. (28), p. 178) and the weight of a directed edge $\pi : G \to \pi G$ is realized by its *step size* $s(\pi) = \log_p(\#\gamma_c G)$ ([2], section 17, eqn. (33), p. 179). These invariants are essential for understanding the phenomenon of *multifurcation* in Definition 2.7. In particular, we can hide multifurcation by restricting all edges π to step size $s(\pi) = 1$, that is, by considering the minimal tree $\mathcal{T}_1(v)$ instead of the entire descendant tree $\mathcal{T}(v)$ of a vertex $v \in V$. In our concrete p-group theoretic model, all vertices G of a minimal tree share a common *coclass*, which is the additive complement $\mathrm{cc}(G) := \mathrm{lo}(G) - \mathrm{cl}(G)$ of the (nilpotency) class $c = \mathrm{cl}(G)$ with respect to the *logarithmic order* $\mathrm{lo}(G) := \log_p(\mathrm{ord}(G))$ of G. Generally, the logarithmic order of an immediate descendant G with parent πG increases by the step size, $\mathrm{lo}(G) = \mathrm{lo}(\pi G) + s(\pi)$, since $\log_p(\#\pi G) = \log_p(\#(G/\ker\pi)) = \log_p(\#G) - \log_p(\#\ker\pi)$. Consequently, the coclass remains fixed in a minimal tree with $s(\pi) = 1$, since

$$\mathrm{cc}(G) = \mathrm{lo}(G) - \mathrm{cl}(G) = \mathrm{lo}(\pi G) + 1 - (\mathrm{cl}(\pi G) + 1) = \mathrm{lo}(\pi G) - \mathrm{cl}(\pi G) = \mathrm{cc}(\pi G).$$

A minimal tree $\mathcal{T}_1(G)$ which contains a unique infinite mainline is called a *coclass tree*. It is denoted by $\mathcal{T}^{(r)}(G) := \mathcal{T}_1(G)$ when its root G is of coclass $r := \mathrm{cc}(G)$. For further details, see ([2], section 5, p. 164).

In view of the principal goals of this chapter, we must specify our intended situation even more concretely. We put $p := 3$, the smallest odd prime number, and we select as the root either $R := \langle 243, 6\rangle$ or $R := \langle 243, 8\rangle$, characterized by its SmallGroup identifier [3]. These are metabelian 3-groups of order $\#R = 243 = 3^5$, logarithmic order $\mathrm{lo}(R) = 5$, class $c = 3$, and coclass $r = 2$.

3.1. Periodicity of finite patterns

Within the frame of the above-mentioned model with $p = 3$ for the theory of rooted in-trees as developed in Section 2, the following finiteness and periodicity statement becomes provable.

The *virtual periodicity* of depth-pruned branches of coclass trees has been proven rigorously with analytic methods (using zeta functions and cone integrals) by du Sautoy [4] in 2000, and with algebraic methods (using cohomology groups) by Eick and Leedham-Green [5] in 2008. We recall that a coclass tree contains a unique infinite path of edges π with uniform step size $s(\pi) = 1$, the so-called mainline. Pattern recognition and pattern classification concern the branches.

Theorem 3.1. *(A finite periodically repeating pattern.)*

*Among the vertices of any mainline $(v_i)_{i\geq 0}$ in $\mathcal{T}(R)$, there exists a **periodic root** v_ϱ with $\varrho \geq 0$ and a **period length** $\lambda \geq 1$ such that the branches*

$$\mathcal{B}(v_{i+\lambda}) \simeq \mathcal{B}(v_i)$$

*are isomorphic **finite** graphs, for all $i \geq \varrho$. Up to a finite preperiodic component, the minimal tree $\mathcal{T}_1(v_0)$ consists of periodically repeating copies of the finite pattern $\dot{\bigcup}_{i=0}^{\lambda-1} \mathcal{B}(v_{\varrho+i})$.*

Proof. According to [4, 5], the claims are true for pruned branches with any fixed depth. However, for $p = 3$ and under the pruning operation on $\mathcal{T}(R)$ described in Section 3.2.2, the virtual periodicity becomes a strict periodicity, since the depth is bounded uniformly for all branches. □

Before we visualize a particular instance of Theorem 3.1 in the diagram of **Figure 1**, we have to establish techniques for disentangling dense branches of high complexity.

3.2. Graph dissection by independent component analysis

3.2.1. Dissection by Galois action

Figure 1 visualizes a graph dissection of the tree $\mathcal{T}(R)$ by independent component analysis. This technique drastically reduces the complexity of visual representations and avoids overlaps of dense subgraphs. The left hand scale gives the order of groups whose isomorphism classes are represented by vertices of the graph. The mainline skeleton (black) connects branches of non σ-groups (red) in the left subfigure and branches of σ-groups (green) in the right subfigure. This terminology has its origin in the action of the Galois group $\mathrm{Gal}(F/\mathbb{Q})$ on the abelianization $\mathfrak{M}/\mathfrak{M}'$, when a vertex of $\mathcal{T}(R)$ is realized as second 3-class group $\mathfrak{M} := \mathrm{Gal}\left(F_3^{(2)}/F\right)$ of an algebraic number field F. For quadratic fields F, we obtain σ-groups.

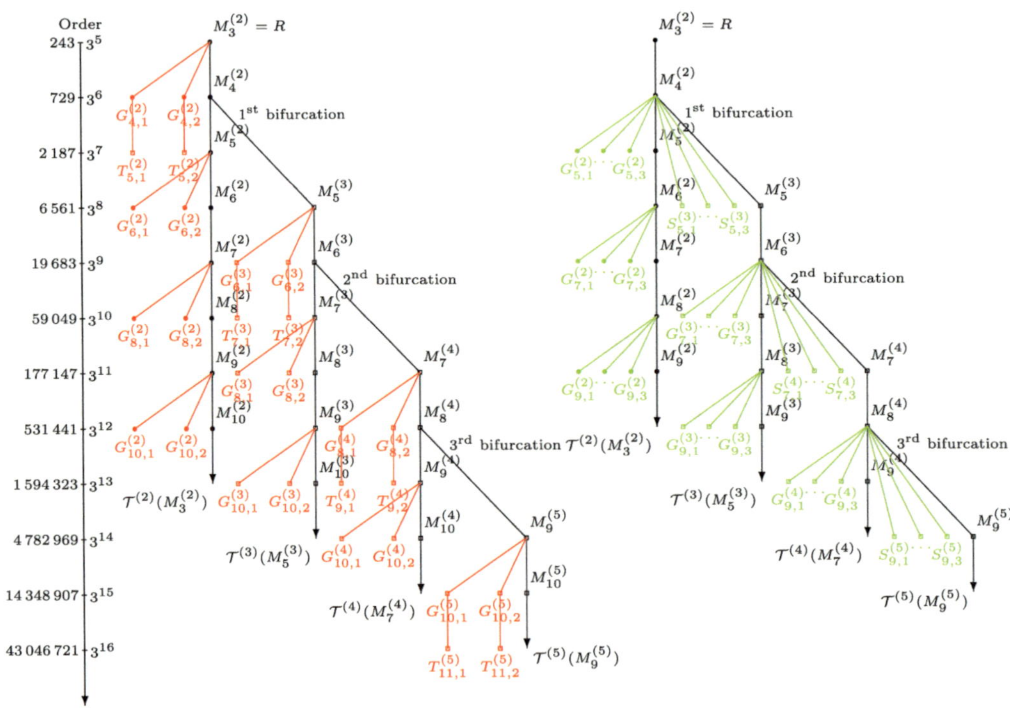

Figure 1. Graph dissection into pruned branches connected by the mainline skeleton.

Definition 3.1. A *σ-group* G admits an automorphism $\sigma \in \mathrm{Aut}(G)$ acting as inversion $\sigma(x) = x^{-1}$ on the commutator quotient G/G'.

The actual graph $\mathcal{T}(R)$ consists of the overlay (superposition) of both subfigures in **Figure 1**. Infinite mainlines are indicated by arrows. The periodic bifurcations form an infinite path with edges of alternating step sizes 1 and 2, according to Theorem 3.2. We call it the *maintrunk*.

With the aid of **Figure 1**, a particular instance of Theorem 3.1 can be expressed in a more concrete and ostensive way by taking the tree root as the ending vertex $v_0 := R$ of the mainline $(v_i)_{i \geq 0}$, and by using the variable class $c \geq 3$ and the fixed coclass $r = 2$ as parameters describing all mainline vertices $M_c^{(r)} := v_{c-3}$. The periodic root is $M_5^{(2)} = v_2$ with $\varrho = 2$ and the period length is $\lambda = 2$. The finite periodic pattern consists of the two branches $\mathcal{B}(M_5^{(2)}) = \{M_5^{(2)}, G_{6,1}^{(2)}, G_{6,2}^{(2)}\}$ (red) and $\mathcal{B}(M_6^{(2)}) = \{M_6^{(2)}, G_{7,1}^{(2)}, G_{7,2}^{(2)}, G_{7,3}^{(2)}\}$ (green). The preperiod is irregular and consists of the two branches $\mathcal{B}(M_3^{(2)}) = \{M_3^{(2)}, G_{4,1}^{(2)}, G_{4,2}^{(2)}, T_{5,1}^{(2)}, T_{5,2}^{(2)}\}$ (red) and $\mathcal{B}(M_4^{(2)}) = \{M_4^{(2)}, G_{5,1}^{(2)}, G_{5,2}^{(2)}, G_{5,3}^{(2)}\}$ (green). But $M_4^{(2)}$ is not coclass-settled, has nuclear rank $n = 2$ and gives rise to a bifurcation with immediate descendants $S_{5,1}^{(3)}, S_{5,2}^{(3)}, S_{5,3}^{(3)}, M_5^{(3)}$ (green) of step size $s = 2$.

3.2.2. *Dissection by Artin transfers*

In **Figure 1**, we have tacitly used a second technique of graph dissection by independent component analysis. **Figure 2** is restricted to the coclass tree $\mathcal{T}^{(2)}(R)$ with exemplary root $R = \langle 243, 8 \rangle$, which is the leftmost coclass tree in both subfigures of **Figure 1**. However, now this coclass tree is drawn completely up to logarithmic order 15, containing both, nonσ-branches and σ-branches. The tree is embedded in a kind of coordinate system having the transfer kernel type (TKT) $\varkappa$ as its horizontal axis and the first component $\tau(1)$ of the transfer target type (TTT) τ as its vertical axis ([6], Dfn. 4.2, p. 27). It is convenient to employ a second graph dissection, according to three fundamental types of transfer kernels

- the vertices with *simple* types E.8, $\varkappa = (1231)$, and E.9, $\varkappa = (2231)$, which are leaves (left of the mainline), except those of order 3^6,
- the vertices with *scaffold* (or skeleton) type c.21, $\varkappa = (0231)$, which are either infinitely capable mainline vertices or nonmetabelian leaves (immediately right of the mainline),
- the vertices with *complex* type G.16, $\varkappa = (4231)$, which are capable at depth 1 and give rise to a complicated brushwood of various descendants (right of the mainline).

The tacit omission in **Figure 1** concerns all vertices with complex type and the leaves with scaffold type. Our main results in this chapter will shed complete light on all mainline vertices and the vertices with simple types. Underlined boldface integers in **Figure 2** indicate the minimal discriminants d of (real and imaginary) quadratic fields $F = \mathbb{Q}(\sqrt{d})$ whose second 3-class group $G_3^{(2)}F := \mathrm{Gal}\left(F_3^{(2)}/F\right)$ realizes the vertex surrounded by the adjacent oval. Three leaves of type E.8 are drawn with red color, because they will be referred to in Theorem 4.1 on 3-class towers.

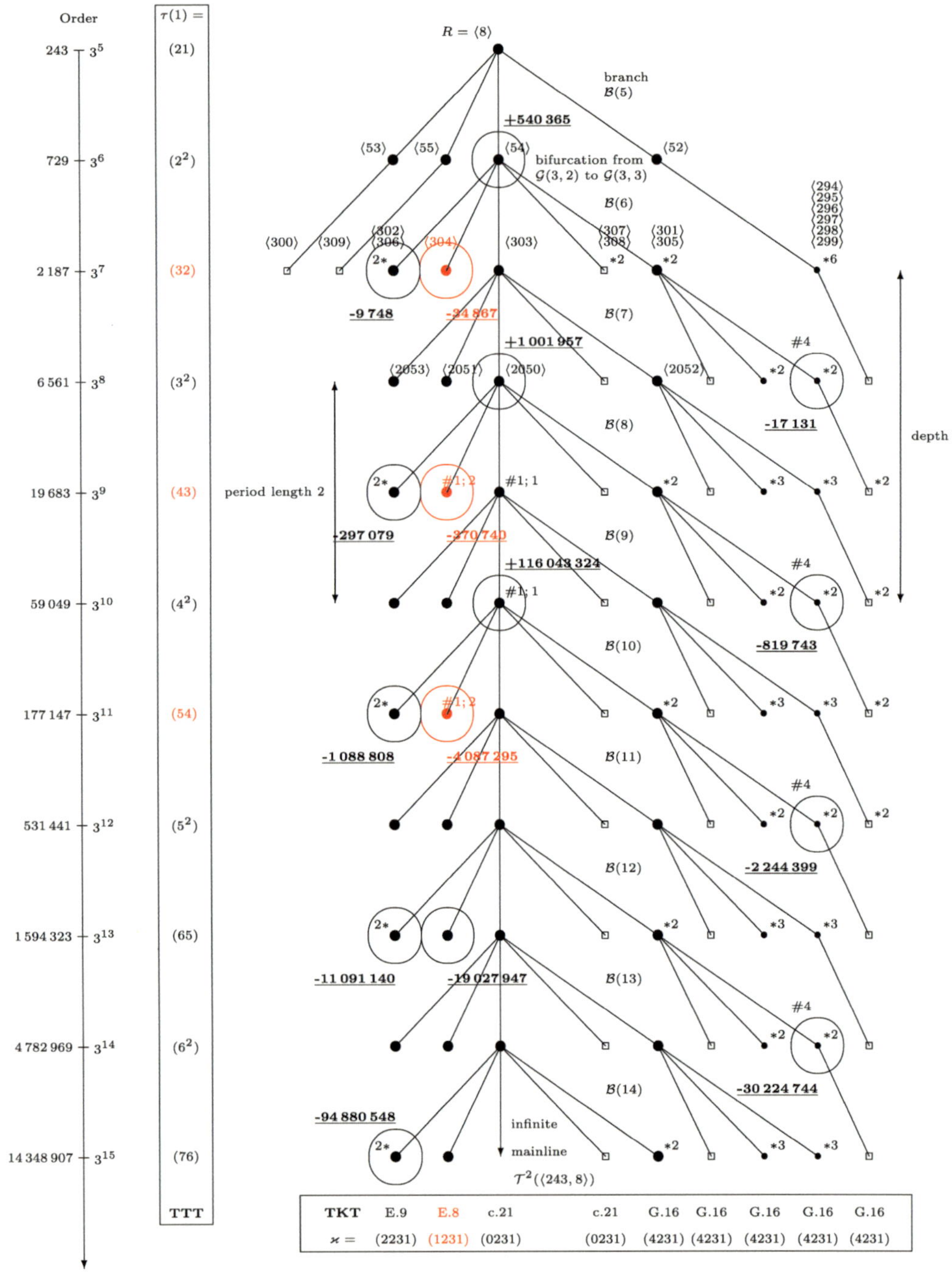

Figure 2. Coclass tree $\mathcal{T}^{(2)}(\langle 243, 8\rangle)$ with simple, scaffold, and complex types.

3.3. Periodicity of infinite patterns

With the aid of a combination of top down and bottom up techniques, we are now going to provide evidence of a new kind of *periodic bifurcations* in pruned descendant trees which contain a unique infinite path of edges π with strictly alternating step sizes $s(\pi) = 1$ and $s(\pi) = 2$, the so-called *maintrunk*. It is very important that the trees are *pruned* in the sense explained at the end of the preceding Section 3.2.2; for otherwise, the maintrunk will not be unique. In fact, each of our pruned descendant trees $\mathcal{T}(R)$ is a countable disjoint union of pruned coclass trees $\mathcal{T}^{(r)}$, $r \geq 2$, which are isomorphic as infinite graphs and connected by edges of weight 2, and finite batches $\mathcal{T}_0^{(r)}$, $r \geq 3$, of sporadic vertices outside of coclass trees. The top down and bottom up techniques are implemented simultaneously in two recursive Algorithms 3.1 and 3.2.

The first Algorithm 3.1 recursively constructs the mainline vertices $M_c^{(r)}$, with class $c \geq 2r-1$, of the coclass tree $\mathcal{T}^{(r)} \subset \mathcal{T}(R)$, for an assigned value $r \geq 2$, by means of the *bottom up* technique. In each recursion step, the *top down* technique is used for constructing the class-c quotient $\mathcal{L}_c^{(r)}$ of an infinite limit group $\mathcal{L}^{(r)}$. Finally, the isomorphism $M_c^{(r)} \simeq \mathcal{L}_c^{(r)}$ is proved.

Theorem 3.2 *(An infinite periodically repeating pattern.) Let $u_r = 30$ be an upper bound. An infinite path is generated recursively, since for each $2 \leq r < u_r$, the immediate descendant $M_{2r+1}^{(r+1)} = M_{2r}^{(r)} - \#2; 1$ of step size 2 of the second mainline vertex $M_{2r}^{(r)}$ of the coclass tree $\mathcal{T}^{(r)}(M_{2r-1}^{(r)})$ is root of a new coclass tree $\mathcal{T}^{(r+1)}(M_{2(r+1)-1}^{(r+1)})$. The pruned coclass trees*

$$\mathcal{T}^{(r)}(M_{2r-1}^{(r)}) \simeq \mathcal{T}^{(2)}(M_3^{(2)})$$

are isomorphic **infinite** *graphs, for each $2 \leq r \leq u_r$. Note that the nuclear rank $n(M_{2r}^{(r)}) = 2$.*

This is the first main theorem of the present chapter. The proof will be conducted with the aid of an infinite limit group $\mathcal{L}_\pm$, due to M. F. Newman. Certain quotients of $\mathcal{L}_\pm$ give precisely the mainline vertices $M_c^{(r)}$ with $r \geq 2$ and $c \geq 2r-1$ as will be shown in Theorem 3.3 and Remark 3.2.

Conjecture 3.1. Theorem 3.2 remains true for any upper bound $u_r > 30$.

3.4. Mainlines of the pruned descendant tree $\mathcal{T}(R)$

Definition 3.2. The complete theory of the mainlines in $\mathcal{T}(R)$ is based on the group

$$\mathcal{L}_\pm := \langle\, a, t \mid (at)^3 = a^3,\ [[t, a], t] = a^{\pm 3} \,|\rangle. \tag{3}$$

For each $r \geq 2$, quotients of $\mathcal{L}_\pm$ are defined by

$$\mathcal{L}_\pm^{(r)} := \mathcal{L}_\pm \,/\, \langle\, a^{3^r} \,\rangle. \tag{4}$$

For each $r \geq 2$, and for each $c \geq 2r - 1$, quotients of $\mathcal{L}_{\pm}^{(r)}$ are defined by

$$\mathcal{L}_{\pm,c}^{(r)} := \begin{cases} \mathcal{L}_{\pm}^{(r)} / \langle [t,a]^{3^\ell} \rangle & \text{if } c = 2\ell + 1 \text{ odd, } \ell \geq r - 1, \\ \mathcal{L}_{\pm}^{(r)} / \langle t^{3^\ell} \rangle & \text{if } c = 2\ell \text{ even, } \ell \geq r. \end{cases} \tag{5}$$

The following Algorithm 3.1 is based on iterated applications of the p-group generation algorithm by Newman [7] and O'Brien [8]. It starts with the root R, given by its compact presentation, and constructs an initial section of the unique infinite maintrunk with strictly alternating step sizes 1 and 2 in the pruned descendant tree $\mathcal{T}(R)$. In each step, the required selection of the child with appropriate transfer kernel type (TKT) is achieved with the aid of our own subroutine IsAdmissible(), which is an elaborate version of ([9], section 4.1, p. 76). After reaching an assigned coclass r = hb+2, our algorithm navigates along the mainline of the coclass tree $\mathcal{T}^{(r)} \subset \mathcal{T}(R)$ and tests each vertex for isomorphism to the corresponding quotient $\mathcal{L}_{\pm,c}^{(r)}$ of class $c \leq 2r - 1 + vb$.

Algorithm 3.1. (Mainline vertices.)

Input: prime p, compact presentation cp of the root, bounds hb, vb, sign s.

Code: uses the subroutine `IsAdmissible()`.

```
r := 2; // initial coclass
Root := PCGroup(cp);
for i in[ 1..hb] do // bottom up in double steps along the maintrunk
    Des := Descendants(Root,NilpotencyClass(Root)+1: StepSizes:=[ 1] );
    for j in[ 1..#Des] do
        if IsAdmissible(Des[ j] ,p,0) then
            Root := Des[ j] ;
        end if;
    end for;
    r := r + 1; // coclass recursion
    Des := Descendants(Root,NilpotencyClass(Root)+1: StepSizes:=[ 2] );
    for j in[ 1..#Des] do
        if IsAdmissible(Des[ j] ,p,0) then
            Root := Des[ j] ;
        end if;
    end for;
end for;
c := 2*r - 1; // starting class c in dependence on the coclass r
er := p^r; l := (c - 1) div 2; ec := p^l;
M<a,t> := Group<a,t| (a*t)^p=a^p, ((t,a),t)=a^(s*p),a^er=1,(t,a)^ec=1>;
QM,pM := pQuotient(M,p,c); // top down construction
if IsIsomorphic(Root,QM) then // identification
    printf "Isomorphism for cc=%o, cl=%o.\n",r,c;
end if;
```

```
for i in[ 1..vb] do // bottom up in single steps along a mainline
    c := c + 1; // nilpotency class recursion
    if (0 eq c mod 2) then // even nilpotency class
        l := c div 2; ec := p^l;
        M<a,t> := Group<a,t| (a*t)^p=a^p, ((t,a),t)=a^(s*p),a^er=1,t^ec=1>;
    else // odd nilpotency class
        l := (c - 1) div 2; ec := p^l;
        M<a,t>:= Group<a,t| (a*t)^p=a^p,((t,a),t)=a^(s*p),a^er=1,(t,a)^ec=1>;
    end if;
    QM,pM := pQuotient(M,p,c); // top down construction
    Des := Descendants(Root,NilpotencyClass(Root)+1: StepSizes:=[ 1] );
    for j in[ 1..#Des] do
        if IsAdmissible(Des[ j] ,p,0) then
            Root := Des[ j] ;
        end if;
    end for;
    if IsIsomorphic(Root,QM) then // identification
        printf "Isomorphism for cc=%o, cl=%o.\n",r,c;
    end if;
end for;
```

Output: coclass `r` and class `c` in each case of an isomorphism.

Remark 3.1. Algorithm 3.1 is designed to be called with input parameters the prime `p=3` and `cp` the compact presentation of either the root $\langle 243, 6\rangle$ with sign `s=-1` or the root $\langle 243, 8\rangle$ with sign `s=+1`. In the current version V2.22-7 of the computational algebra system MAGMA [10], the bounds are restricted to `r=hb+2≤8` and `c=vb+2r−1≤35`, since otherwise the maximal possible internal word length of relators in MAGMA is surpassed. Close to these limits, the required random access memory increases to a considerable value of approximately 8 GB RAM.

Theorem 3.3. *(Mainline vertices as quotients of the limit group $\mathcal{L}_{\pm}$.) Let $u_r := 8$, $u_c := 35$.*

1. *For each $2 \le r \le u_r$, and for each $2r-1 \le c \le u_c$, the mainline vertex $M_c^{(r)}$ of coclass r and nilpotency class c in the tree $\mathcal{T}(R)$ is isomorphic to $\mathcal{L}_{\pm,c}^{(r)}$.*
2. *For each $2 \le r \le u_r$, the projective limit of the mainline $\left(M_c^{(r)}\right)_{c \ge 2r-1}$ with vertices of coclass r in the tree $\mathcal{T}(R)$ is isomorphic to $\mathcal{L}_{\pm}^{(r)}$.*
3. *$\mathcal{L}_{\pm}$ is an infinite nonnilpotent profinite limit group.*

Proof. (1) The repeated execution of Algorithm 3.1 for successive values from `hb:=0` to `hb:=6`, with input data `p:=3`, `cp:=CompactPresentation(SmallGroup(243,i))`, $i \in \{6, 8\}$, $s \in \{-1, +1\}$, and `vb:=32`, proves the isomorphisms $M_c^{(r)} \simeq \mathcal{L}_{\pm,c}^{(r)}$ for $2 \le r \le u_r = 8$ and $2r - 1 \le c \le u_c = 35$. The algorithm is initialized by the starting group $R = M_3^{(2)} = \langle 243, i\rangle$ of coclass `r:=2`. The first loop moves along the maintrunk recursively with strictly alternating step sizes 1 and 2 until the root $M^{(r)}_{2r-1}$ of the coclass tree $\mathcal{T}^{(r)}$ with `r=2+hb` is reached. The second loop iterates through the mainline vertices $M^{(r)}_c$, $c \geq 2r-1$, of the coclass tree $\mathcal{T}^{(r)}\left(M^{(r)}_{2r-1}\right)$, always checking for isomorphism to the appropriate quotient $\mathcal{L}^{(r)}_{\pm,c}$. The subroutine `IsAdmissible()` tests the transfer kernel type of all descendants and selects the unique capable descendant with type c.18, respectively, c.21. (2) Since periodicity sets in for $2u_r - 1 = 17 \leq c \leq u_c = 35$, the claim is a consequence of Theorem 3.1. (3) The quotient $\mathcal{L}^{(1)}_{\pm}$ is already infinite and nonnilpotent. Adding the relation $[t, t^a, t] = 1$ suffices to give $[t, a, t]$ central and $\mathcal{L}_{\pm}$ profinite. □

Conjecture 3.2. Theorem 3.3 remains true for arbitrary upper bounds $u_r > 8$, $u_c > 35$.

Remark 3.2. When the top down constructions in Algorithm 3.1 are cancelled, the bottom up operations are still able to establish much bigger initial sections of the infinite maintrunk and of the infinite coclass tree with fixed coclass $r \geq 2$. Admitting an increasing amount of CPU time, we can easily reach astronomic values of the coclass, $r = 32$, and the nilpotency class, $c = 63$, that is a logarithmic order of $r + c = 95$, without surpassing any internal limitations of MAGMA, and the required storage capacity remains quite modest, i.e., clearly below 1 GB RAM. This remarkable stability underpins Conjecture 3.2 with additional support from the bottom up point of view.

3.5. Covers of metabelian 3-groups

Only one of the coclass subtrees $\mathcal{T}^{(r)}$, $r \geq 2$, of the entire rooted in-tree $\mathcal{T}(R)$ contains metabelian vertices, namely the first subtree $\mathcal{T}^{(2)}$. The following theorem shows how transfer kernel types are distributed among metabelian vertices G of depth $\mathrm{dp}(G) \leq 1$ on the tree $\mathcal{T}^{(2)}$, as partially illustrated by the **Figures 1** and **2**.

Theorem 3.4. *(Metabelian vertices of the coclass tree* $\mathcal{T}^{(2)}R$*.)*

For each finite 3-group G*, we denote by* $c := \mathrm{cl}(G)$ *the nilpotency class, by* $r := \mathrm{cc}(G)$ *the coclass, and by* $\varkappa$ *the transfer kernel type of* G*. More explicitly, such a group is also denoted by* $G = G^{(r)}_c$*. The following statements describe the structure of the metabelian skeleton of the coclass tree* $\mathcal{T}^{(2)}R$ *with root* $R := \langle 243, 6\rangle$*, respectively,* $R := \langle 243, 8\rangle$*, down to depth* 1.

1. *For each* $c \geq 3$*, the mainline vertex* $M^{(2)}_c$ *of the coclass tree possesses type* c.18, $\varkappa = (0122)$*, respectively,* c.21, $\varkappa = (0231)$.

2. *For each* $c \geq 4$*, there exists a unique child* $G^{(2)}_{c,1}$ *of* $M^{(2)}_{c-1}$ *with type* E.6, $\varkappa = (1122)$*, respectively,* E.8, $\varkappa = (1231)$.

3. *For even* $c \geq 4$*, there exists a unique child* $G^{(2)}_{c,2}$ *of* $M^{(2)}_{c-1}$ *with type* E.14, $\varkappa = (3122)$*, respectively,* E.9, $\varkappa = (2231)$*. Thus,* $N_1(M^{(2)}_{c-1}) = 3$ *and* $C_1(M^{(2)}_{c-1}) = 1$*, in the pruned tree.*

4. *For odd* $c \geq 5$*, there exist two children* $G^{(2)}_{c,2}$*,* $G^{(2)}_{c,3}$ *of* $M^{(2)}_{c-1}$ *with type* E.14, $\varkappa = (3122) \sim (4122)$*, respectively,* E.9, $\varkappa = (2231) \sim (3231)$*. Thus,* $N_1(M^{(2)}_{c-1}) = 4$ *and* $C_1(M^{(2)}_{c-1}) = 1$.

5. *For even* $c \geq 4$, *there exists a unique child* $G_{c,4}^{(2)}$ *of* $M_{c-1}^{(2)}$ *with type* H.4, $\varkappa = (2122)$, *respectively,* G.16, $\varkappa = (4231)$. *It is removed from the pruned tree.*

6. *For odd* $c \geq 5$, *there exist two children* $G_{c,4}^{(2)}, G_{c,5}^{(2)}$ *of* $M_{c-1}^{(2)}$ *with type* H.4, $\varkappa = (2122)$, *respectively,* G.16, $\varkappa = (4231)$. *They are removed from the pruned tree.*

Proof. See Nebelung ([11], Lemma 5.2.6, p. 183, Figures, p. 189 f., Satz 6.14, p. 208). □

Definition 3.3. For $e \in \{0, 1\}$, we define the *cover limit,* due to M. F. Newman, to be the group

$$\begin{aligned} \mathcal{C}^{(e)} := \langle\ a, t, u, y, z \mid t^a = u,\ u^a tuy = [u, t]^e,\ a^3[t, a, t] = z,\ [u, t, t] = [u, t, u] = 1, \\ y^3 = 1, [a, y] = [t, y] = [u, y] = [z, y] = 1, z^3 = 1, [t, z] = [u, z] = 1\ \rangle, \end{aligned} \tag{6}$$

which was introduced in [12]. For each $k \in \{-1, 0, 1\}$ and for each integer $c \geq 4$, let

$$\mathcal{Q}_c^{(e,k)} := \mathcal{C}^{(e)} \ / \ \langle\ y w_c^k v_c,\ z w_c\ \rangle \tag{7}$$

be the *class-c quotient with parameter k* of $\mathcal{C}^{(e)}$, where $w_c := [t, \overbrace{a, \ldots, a}^{(c-1)\text{times}}]$ and $v_c := [w_{c-2}, [t, a]]$.

In each step, $i \geq 1$, of the second Algorithm 3.2, the *top down* technique constructs a certain class-c quotient $\mathcal{Q}_c$, $c = i + 3$, of a fixed infinite pro-3 group $\mathcal{C}$, the *cover limit,* and the *bottom up* technique constructs all metabelian children of a certain vertex M_{i-1} on the mainline of the first coclass tree $\mathcal{T}^{(2)}(R)$, and selects, firstly, the next vertex M_i of depth $\mathrm{dp}(M_i) = 0$ on the mainline of $\mathcal{T}^{(2)}(R)$ for continuing the recursion, secondly, a vertex G_i of depth $\mathrm{dp}(G_i) = 1$ with assigned transfer kernel type $\varkappa(G_i)$. Each recursion step is completed by proving that G_i is isomorphic to the second derived quotient $\mathcal{Q}_c/\mathcal{Q}''_c$, that is, $\mathcal{Q}_c \in \mathrm{cov}(G_i)$ belongs to the *cover* of G_i in the sense of ([13], section 1.3, Dfn. 1.1, p. 75). More precisely, we have $M_i = M_{i+3}^{(2)}$ and $G_i = G_{i+3,j}^{(2)}$ with some j.

Algorithm 3.2. (Shafarevich cover.)

Input: prime p, compact presentation cp of the root, bound vb, parameters e and k.

Code: uses the subroutine `IsAdmissible()`.

```
C<a,t,u,y,z> := Group< a,t,u,y,z |
     y^p, (a,y), (t,y), (u,y), (y,z), (t,z), (u,z), z^p,
     (u,t,t), (u,t,u), t^a = u, u^a*t*u*y*(u,t)^-e, a^p*(t,a,t) = z>;
Root := PCGroup(cp);
Leaf := Root;
for i in[1..vb] do // bottom up along the mainline of coclass 2
    c := i + 3; // nilpotency class
    w :=[t];
    for j in[1..c] do // construction of iterated commutator
        s := (w[j],a);
```

```
        Append(~w,s);
    end for;
    w1 := w[c-2]^-1*(a,t)*w[c-2]*(t,a);
    H := quo<C | y*w[c]^k*w1, z*w[c]>;
    Q,pQ := pQuotient(H,p,c); // top down construction of Shafarevich cover
    Des := Descendants(Root,NilpotencyClass(Root)+1);
    m := 0;
    for cnt in[1..#Des] do
        if IsAdmissible(Des[cnt],p,0) then
            Root := Des[cnt]; // next mainline vertex
        elif IsAdmissible(Des[cnt],p,2) then
            m := m+1;
            if (1 eq m) then
                Leaf := Des[cnt]; // first leaf with assigned TKT
            end if;
        end if;
    end for;
    DQ := DerivedSubgroup(Q);
    D2Q := DerivedSubgroup(DQ);
    Q2Q := Q/D2Q; // metabelianization
    if IsIsomorphic(Leaf,Q2Q) then // identification
        printf "Dsc.cl.%o isomorphic to 2nd drv.qtn. of Cov.cl.%o.\n",c,c;
    end if;
end for;
```

Output: nilpotency class c in each case of an isomorphism.

The next theorem is the second main result of this chapter, establishing the finiteness and structure of the *cover* for each metabelian 3-group with transfer kernel of type E.

Theorem 3.5. *(Explicit covers of metabelian 3-groups.) Let $u := 8$ be an upper bound and $G^{(2)}_{c,j}$ in $\mathcal{T}^{(2)}\left(M^{(2)}_3\right)$ be the metabelian 3-group of nilpotency class $c \geq 4$ with transfer kernel type*

$$\varkappa = \begin{cases} (1122),\ \text{E.6, resp. } (1231),\ \text{E.8} & \text{if } j = 1, \\ (3122),\ \text{E.14, resp. } (2231),\ \text{E.9} & \text{if } j = 2 \text{ or } (j = 3 \text{ and } c \text{ odd}). \end{cases}$$

1. *The* **cover** *of $G^{(2)}_{c,j}$ is given by*

$$\operatorname{cov}\left(G^{(2)}_{c,j}\right) = \begin{cases} \left\{G^{(2)}_{c,j}; G^{(3)}_{c,j}, \ldots, G^{(\ell+1)}_{c,j}, G^{(\ell+2)}_{c,j}, T^{(\ell+2)}_{c+1,j}\right\} & \text{if } c = 2\ell+4,\ 1 \leq j \leq 2, \\ \left\{G^{(2)}_{c,j}; G^{(3)}_{c,j}, \ldots, G^{(\ell+1)}_{c,j}, G^{(\ell+2)}_{c,j}, S^{(\ell+3)}_{c,j}\right\} & \text{if } c = 2\ell+5,\ 1 \leq j \leq 3. \end{cases} \quad (8)$$

where $0 \leq \ell \leq u$. In particular, the cover is a finite set with $\ell + 2$ elements ($\ell + 1$ of them nontrivial), which are nonσ-groups for even $c \geq 4$, and σ-groups for odd $c \geq 5$.

2. *The* **Shafarevich cover** *of* $G_{c,j}^{(2)}$ *with respect to imaginary quadratic fields F is given by*

$$\mathrm{cov}\left(G_{c,j}^{(2)}, F\right) = \begin{cases} \emptyset & \text{if } c = 2\ell + 4,\ 0 \le \ell \le u,\ 1 \le j \le 2, \\ \left\{S_{c,j}^{(\ell+3)}\right\} & \text{if } c = 2\ell + 5,\ 0 \le \ell \le u,\ 1 \le j \le 3. \end{cases} \tag{9}$$

In particular, the Shafarevich cover contains a unique Schur σ-group, if $c \ge 5$ *is odd.*

3. *The class-c quotient with parameter k of the cover limit* $\mathcal{C}^{(e)}$ *is isomorphic to a Schur σ-group* $\mathcal{Q}_c^{(e,k)} \simeq S_{c,j}^{(\ell+3)}$, *for* $c = 2\ell + 5$ *or to a nonσ-group* $\mathcal{Q}_c^{(e,k)} \simeq G_{c,j}^{(\ell+2)}$, *for* $c = 2\ell + 4$. *The precise correspondence between the parameters k and j is given in the following way.*

$$\text{Types E.6,E.8}: \ \mathcal{Q}_c^{(e,0)} \simeq \begin{cases} S_{c,1}^{(\ell+3)} & \text{for odd class } c = 2\ell + 5,\ 0 \le \ell \le u, \\ G_{c,1}^{(\ell+2)} & \text{for even class } c = 2\ell + 4,\ 0 \le \ell \le u, \end{cases}$$

$$\text{type E.9}: \ \mathcal{Q}_c^{(+1,-1)} \simeq \begin{cases} S_{c,2}^{(\ell+3)} & \text{for odd class } c = 2\ell + 5,\ 0 \le \ell \le u, \\ G_{c,2}^{(\ell+2)} & \text{for even class } c = 2\ell + 4,\ 0 \le \ell \le u, \end{cases}$$

$$\text{type E.9}: \ \mathcal{Q}_c^{(+1,+1)} \simeq \begin{cases} S_{c,3}^{(\ell+3)} & \text{for odd class } c = 2\ell + 5,\ 0 \le \ell \le u, \\ G_{c,2}^{(\ell+2)} & \text{for even class } c = 2\ell + 4,\ 0 \le \ell \le u. \end{cases} \tag{10}$$

In particular, $\mathcal{Q}_c^{(+1,-1)} \simeq \mathcal{Q}_c^{(+1,+1)}$ *for even class* $c = 2\ell + 4,\ 0 \le \ell \le u$.

The variant $e = 0$, *respectively,* $e = 1$, *is associated to the root* $R = \langle 243, 6 \rangle$, *respectively,* $R = \langle 243, 8 \rangle$.

4. *A parameterized family of* **fork topologies** *for second 3-class groups* $\mathrm{Gal}\left(F_3^{(2)}/F\right)$ *of imaginary quadratic fields F is given uniformly for the states* $\uparrow^\ell$ *(ground state for* $\ell = 0$, *excited state for* $1 \le \ell \le u$*) of transfer kernels with type* E *by the symmetric topology symbol*

$$P = \overbrace{\mathrm{E} \xrightarrow{1}}^{\text{Leaf}} \quad \overbrace{\left\{\mathrm{c} \xrightarrow{1}\right\}^{2\ell}}^{\text{Mainline}} \quad \overbrace{\mathrm{c}}^{\text{Fork}} \quad \overbrace{\left\{\xleftarrow{2} \mathrm{c} \xleftarrow{1} \mathrm{c}\right\}^{\ell}}^{\text{Maintrunk}} \quad \overbrace{\xleftarrow{2} \mathrm{E}}^{\text{Leaf}} \tag{11}$$

with scaffold type c *and the following invariants:*

distance $d = 4\ell + 2$ *(Definition 2.3), weighted distance* $w = 5\ell + 3$ *(Definition 2.4),*

class increment $\Delta\mathrm{cl} = (2\ell + 5) - (2\ell + 5) = 0$, *coclass increment* $\Delta\mathrm{cc} = (\ell + 3) - 2 = \ell + 1$,

logarithmic order increment $\Delta\mathrm{lo} = (3\ell + 8) - (2\ell + 7) = \ell + 1$ ([13], Dfn. 5.1, p. 89).

Proof. We compare the uniform generator rank $d_1 = 2$ of all involved groups $G_{c,j}^{(r)}$, $c \ge 4$, $r \ge 2$, $1 \le j \le 3$, with their relation rank d_2. Since $d_2 = \mu$ and the p-multiplicator rank is $\mu = 2$ for $S_{c,j}^{(r)}$

with odd $c = 2\ell + 5 \geq 5$ and $r = \ell + 3 \geq 3$, but $\mu = 3$ otherwise, only the groups $S_{c,j}^{(r)}$ are Schur σ-groups with balanced presentation $d_2 = 2 = d_1$, and are therefore admissible as 3-tower groups of imaginary quadratic fields F, according to our corrected version ([6], section 5, Thm. 5.1, pp. 28–29) of the Shafarevich Theorem ([14], Thm. 6, (18′)). Finally, we remark that the nuclear rank is $\nu = 1$ for $G_{c,j}^{(r)}$ with even $c = 2\ell + 4$, $r = \ell + 2$, and child $T_{c+1,j}^{(r)}$, but $\nu = 0$ otherwise.

The execution of Algorithm 3.2 with input data `p:=3, vb:=25`, either `i:=6, e:=0`, or `i:=8, e:=1`, and `cp:=CompactPresentation(SmallGroup(243,i))`, proves the isomorphisms $\mathcal{Q}_c^{(e,k)} \simeq S_{c,j}^{(\ell+3)}$, $c = 2\ell + 5$, respectively, $\mathcal{Q}_c^{(e,k)} \simeq G_{c,j}^{(\ell+2)}$, $c = 2\ell + 4$, for $4 \leq c \leq 20$, that is, $0 \leq \ell \leq u = 8$. The algorithm is initialized by the starting group $R = M_3^{(2)}$ of coclass 2. The loop navigates through the mainline vertices $M_c^{(2)}$, $c \geq 3$, of the coclass tree $\mathcal{T}^{(2)}\left(M_3^{(2)}\right)$. The subroutine `IsAdmissible()` tests the transfer kernel type of all descendants and selects either the unique capable descendant with type c.18, respectively, c.21, for the flag 0 or the unique descendant with type E.6, respectively, E.8, for the flag 1, or the first or second descendant with type E.9, for the flag 2. The selected nonmainline vertex is always checked for isomorphism to the metabelianization of the appropriate quotient $\mathcal{Q}_c^{(e,k)}$. See also ([2], section 21.2, pp. 189–193), ([15], pp. 751–756), the proof of Theorem 4.1, and **Figures 5–7**.

Here again, a pure bottom up approach without top down constructions, instead of using Algorithm 3.2, is able to reach coclass $r = 32$, nilpotency class $c = 63$, and logarithmic order $r + c = 95$, without surpassing internal limits of MAGMA, and strongly supports Conjecture 3.3.

Conjecture 3.3. Theorem 3.5 remains true for any upper bound $u > 8$.

Figure 3 shows exactly the same situation as **Figure 1**, supplemented by blue arrows indicating the projections of the quotients $\mathcal{Q}_c^{(e,k)}$ onto their metabelianizations, that is, $S_{c,j}^{(\ell+3)} \to G_{c,j}^{(2)}$, for odd class $c = 2\ell + 5$, in the right diagram with green branches, and $G_{c,j}^{(\ell+2)} \to G_{c,j}^{(2)}$, for even class $c = 2\ell + 4$, in the left diagram with red branches. For $c = 4$, a degeneration occurs, since $\mathcal{Q}_4^{(e,k)}$ is metabelian already, indicated by surrounding blue circles.

Strictly speaking, the caption of **Figure 3**, in its full generality, is valid for $e = 1$, $M_3^{(2)} = \langle 243, 8\rangle$ only. For $e = 0$, $M_3^{(2)} = \langle 243, 6\rangle$, all blue arrows have the same meaning as before but the interpretation of the covers as quotients $\mathcal{Q}_c^{(e,k)}$ is slightly restricted. Whereas we have the following supplement to Eq. (10):

$$\text{type E.14}: \ \mathcal{Q}_c^{(0,-1)} \simeq \begin{cases} S_{c,3}^{(\ell+3)} & \text{for odd class } c = 2\ell + 5,\ 0 \leq \ell \leq u, \\ G_{c,2}^{(\ell+2)} & \text{for even class } c = 2\ell + 4,\ 0 \leq \ell \leq u, \end{cases} \tag{12}$$

the quotients $\mathcal{Q}_c^{(0,+1)}$ lead into a completely different realm, namely the complicated brushwood of the complex transfer kernel type H.4.

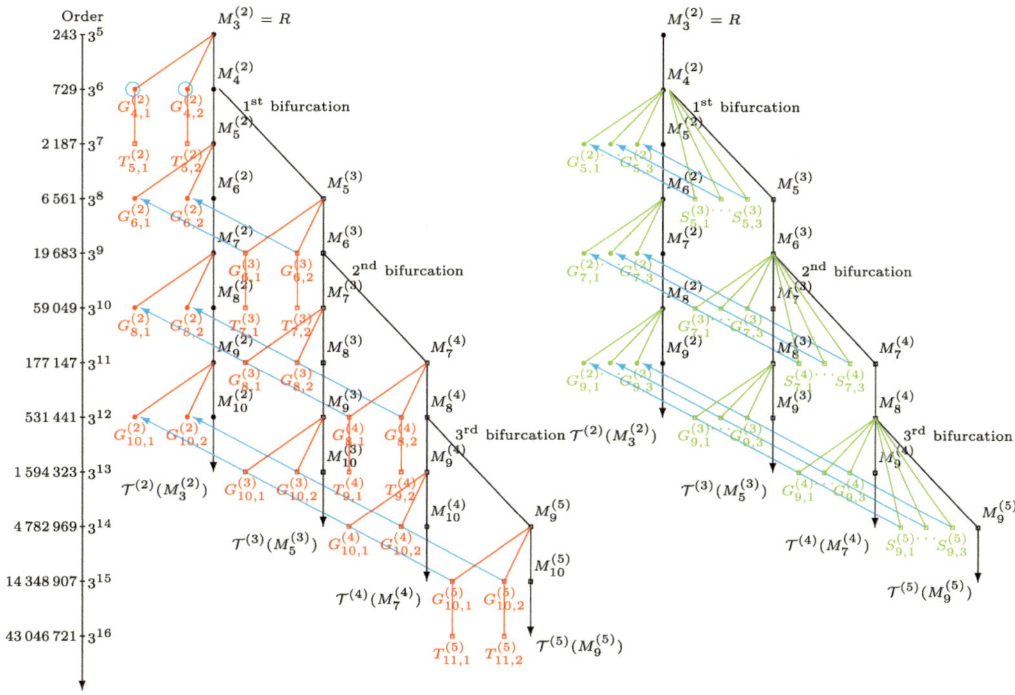

Figure 3. Projections $\mathcal{Q}_c^{(e,k)} \to \mathcal{Q}_c^{(e,k)}/\left(\mathcal{Q}_c^{(e,k)}\right)''$ of the covers onto their metabelianizations.

Figure 4 shows three pruned descendant trees $\mathcal{T}_*(\mathfrak{R})$ with roots $\mathfrak{R} = \langle 243, 4\rangle$, $\mathfrak{R} = \langle 6561, 614\rangle$, and $\mathfrak{R} = \langle 6561, 613\rangle - \#1; 1 - \#2; 1$, all of whose vertices are of type H.4 exclusively. We restrict the trees to σ-groups indicated by green color. The top vertex $\langle 27, 3\rangle$ is intentionally drawn twice to avoid an overlap of the dense trees and to admit a uniform representation of periodic bifurcations.

The tree with root $\langle 243, 4\rangle$ is not concerned by the quotients $\mathcal{Q}_c^{(0,+1)}$. It is sporadic and consists of periodically repeating finite saplings of depth 2 and increasing coclass $2, 3, \ldots$. Connected by the maintrunk with vertices of type c.18 (red color) in the descendant tree $\mathcal{T}(\langle 243, 6\rangle)$, the trees with roots $\langle 6561, 614\rangle$ and $\langle 6561, 613\rangle - \#1; 1 - \#2; 1$ form the beginning of an infinite sequence of similar trees, which are, however, not isomorphic as graphs, since the depth of the constituting saplings increases in steps of 2. The projections of the quotients $\mathcal{Q}_c^{(0,+1)}$ with odd class $c \in \{5, 7\}$ onto their metabelianizations are indicated by blue arrows.

3.6. Topologies in descendant trees

Tree topologies describe the mutual location of distinct higher p-class groups $\mathrm{G}_p^{(m)}F$ and $\mathrm{G}_p^{(n)}F$, with $n > m \geq 1$, of an algebraic number field F. The case $(m, n) = (3, 4)$ will be crucial for finding the first examples of *four-stage towers* of p-class fields with length $\ell_p F := \mathrm{dl}(\mathrm{G}_p^{(\infty)}F) = 4$, which are unknown up to now, for any prime $p \geq 2$. Fork topologies with $(m, n) = (2, 3)$ have proved to be essential for discovering p-class towers with length $\ell_p F = 3$, for odd primes $p \geq 3$.

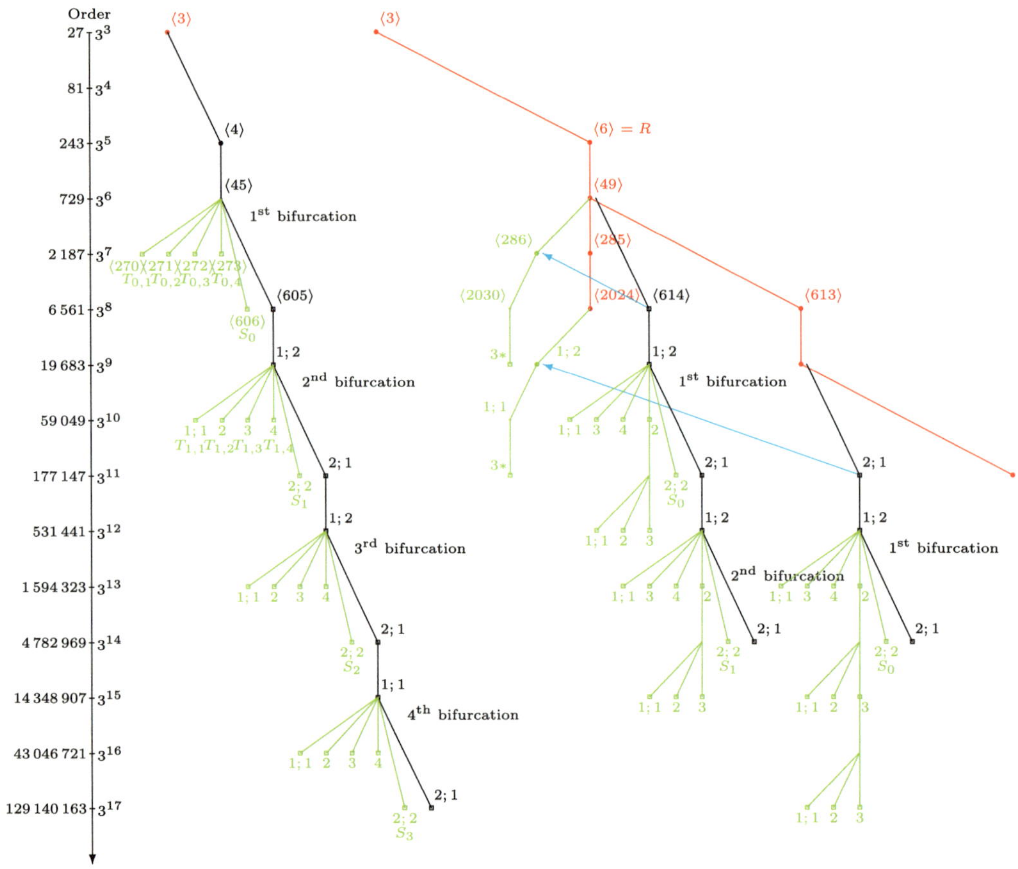

Figure 4. Branches of σ-groups with complex type H.4 connected by the maintrunk.

In ([13], Prp. 5.3, p. 89), we have pointed out that the *qualitative* topology problem for $(m,n) = (1,2)$ is trivial, since the fork of $G_p^{(1)}F$ and $G_p^{(2)}F$ is simply the abelian root $G_p^{(1)}F \simeq Cl_pF$ of the entire relevant descendant tree. However, the *quantitative* structure of the root path between $G_p^{(2)}F$ and $G_p^{(1)}F$ is not at all trivial and can be given in a general theorem for $Cl_pF \simeq (p,p)$ and $p \in \{2,3\}$ only. In the following Theorem 3.6, we establish a purely group theoretic version of this result by replacing $G_p^{(2)}F$ with an arbitrary *metabelian* 3-group $\mathfrak{M}$ having abelianization $\mathfrak{M}/\mathfrak{M}'$ of type $(3,3)$. Any attempt to determine the group $G := \mathrm{Gal}(F_p^{(\infty)}/F)$ of the p-class tower $F_p^{(\infty)}$ of an algebraic number field F begins with a search for the metabelianization $\mathfrak{M} := G/G''$, i.e., the second derived quotient, of the p-tower group G. $\mathfrak{M}$ is also called the *second p-class group* $\mathrm{Gal}(F_p^{(2)}/F)$ of F, and $F_p^{(2)}$ can be viewed as a metabelian approximation of the p-class tower $F_p^{(\infty)}$. In the case of the smallest odd prime $p = 3$ and a number field F with 3-class group Cl_3F of type $(3,3)$, the structure of the root path from $\mathfrak{M}$ to the root $\langle 9,2\rangle$ is known explicitly. For its description, it suffices to use the set of possible transfer kernel types

$$X \in \{A, D, E, F, G, H, a, b, c, d\}$$

of the ancestors $\pi^j\mathfrak{M}$, $0 \le j \le \ell$, and the symbol $\stackrel{s}{\rightarrow}$ for a weighted edge of step size $s \ge 1$ with formal exponents denoting iteration. A capable vertex is indicated by an asterisk X^*.

Theorem 3.6. *(Periodic root paths.)*

There exist basically three kinds of root paths $P := (\pi^j\mathfrak{M})_{0 \le j \le \ell}$ *of metabelian 3-groups* $\mathfrak{M}$ *with abelianization* $\mathfrak{M}/\mathfrak{M}'$ *of type* $(3,3)$*, which are located on coclass trees. Let c denote the nilpotency class* $\mathrm{cl}(\mathfrak{M})$ *and r the coclass* $\mathrm{cc}(\mathfrak{M})$ *of* $\mathfrak{M}$.

1. *If* $r = 1$ *and* $c \ge 1$*, then* $P = X\left\{\stackrel{1}{\rightarrow} a^*\right\}^{c-1}$*, where* $X \in \{A, a, a^*\}$.

2. *If* $r = 2$ *and* $c \ge 3$*, then either* $P = X\{\stackrel{1}{\rightarrow} b^*\}^{c-3} \stackrel{2}{\rightarrow} a^* \stackrel{1}{\rightarrow} a^*$*, where* $X \in \{d, b, b^*\}$*, or* $P = X\{\stackrel{1}{\rightarrow} c^*\}^{c-3} \stackrel{2}{\rightarrow} a^* \stackrel{1}{\rightarrow} a^*$*, where* $X \in \{E, G^*, H^*, c^*\}$*. An additional variant arises for* $r = 2$*,* $c \ge 5$*, with* $P = X \stackrel{1}{\rightarrow} X^*\{\stackrel{1}{\rightarrow} c^*\}^{c-4} \stackrel{2}{\rightarrow} a^* \stackrel{1}{\rightarrow} a^*$*, where* $X \in \{G, H\}$.

3. *If* $r \ge 3$ *and* $c \ge r+1$*, then either* $P = X\{\stackrel{1}{\rightarrow} b^*\}^{c-(r+1)}\{\stackrel{2}{\rightarrow} b^*\}^{r-2} \stackrel{2}{\rightarrow} a^* \stackrel{1}{\rightarrow} a^*$*, where* $X \in \{d, b, b^*\}$*, or* $P = X\{\stackrel{1}{\rightarrow} d^*\}^{c-(r+1)}\{\stackrel{2}{\rightarrow} b^*\}^{r-2} \stackrel{2}{\rightarrow} a^* \stackrel{1}{\rightarrow} a^*$*, where* $X \in \{F, G^*, H^*, d^*\}$*. An additional variant arises for* $r \ge 3$*,* $c \ge r+3$*, with* $P = X \stackrel{1}{\rightarrow} X^*\{\stackrel{1}{\rightarrow} d^*\}^{c-(r+2)}\{\stackrel{2}{\rightarrow} b^*\}^{r-2} \stackrel{2}{\rightarrow} a^* \stackrel{1}{\rightarrow} a^*$*, where* $X \in \{G, H\}$.

In particular, the maximal possible step size is $s = 2$*, and the* $r-1$ *edges with step size* $s = 2$ *arise successively without gaps at the end of the path, except the trailing edge of step size* $s = 1$.

Proof. X always denotes the type of the starting vertex $\mathfrak{M}$. The remaining vertices of the root path form the *scaffold*, which connects the starting vertex with the ending vertex (the root $R = \langle 9, 2\rangle$). The unique coclass tree $\mathcal{T}^{(1)}\langle 9, 2\rangle$ with $r = 1$ has a mainline of type a^*. Two of the coclass trees $\mathcal{T}^{(2)}\langle 243, n\rangle$ with $r = 2$, those with $n \in \{6, 8\}$, have mainlines of type c^* and an additional scaffold of type a^*. For $n = 3$, the mainline is of type b^*. The coclass trees $\mathcal{T}^{(r)}$ with $r \ge 3$ behave uniformly with mainlines of type b^* or d^* and scaffold types b^*, a^*. For details, see Nebelung ([11], Satz 3.3.7, p. 70, Lemma 5.2.6, p. 183, Satz 6.9, p. 202, Satz 6.14, p. 208).

Remark 3.3. The final statement of Theorem 3.6 is a graph theoretic reformulation of the quotient structure of the lower central series $\left(\gamma_j\mathfrak{M}\right)_{j \ge 1}$ of a metabelian 3-group $\mathfrak{M}$, observing that the root $R = \langle 9, 2\rangle$ corresponds to the bicyclic quotient $\gamma_1/\gamma_2 \simeq (3, 3)$ and the conspicuous trailing edge $\stackrel{1}{\rightarrow} a^*$ corresponds to the cyclic bottleneck $\gamma_2/\gamma_3 \simeq (3)$. The structure is drawn ostensibly in eqn. (2.12) of ([16], section 2.2), using the CF-invariant $e = r + 1$ instead of the coclass r.

Theorem 3.6 concerns periodic vertices on coclass trees. Sporadic vertices outside of coclass trees must be treated separately in Corollary 3.1.

Corollary 3.1. *(Sporadic root paths.)*

As before, let $\mathfrak{M}$ *be a metabelian* $_3$*-group with abelianization* $\mathfrak{M}/\mathfrak{M}' \simeq (3,3)$*, nilpotency class* $c := \mathrm{cl}(\mathfrak{M})$*, and coclass* $r := \mathrm{cc}(\mathfrak{M})$*. Assume that* $\mathfrak{M}$ *is located outside of coclass trees.*

1. *If* $r = 2$ *and* $c = 3$*, then* $P = \mathrm{X} \xrightarrow{2} \mathrm{a}^* \xrightarrow{1} \mathrm{a}^*$*, where* $\mathrm{X} \in \{\mathrm{D}, \mathrm{G}^*, \mathrm{H}^*\}$.

2. *If* $r = 2$ *and* $c = 4$*, then* $P = \mathrm{X} \xrightarrow{1} \mathrm{X}^* \xrightarrow{2} \mathrm{a}^* \xrightarrow{1} \mathrm{a}^*$*, where* $\mathrm{X} \in \{\mathrm{G}, \mathrm{H}\}$.

3. *If* $r \geq 3$ *and* $c = r+1$*, then* $P = \mathrm{X}\left\{\xrightarrow{2} \mathrm{b}^*\right\}^{r-2} \xrightarrow{2} \mathrm{a}^* \xrightarrow{1} \mathrm{a}^*$*, where* $\mathrm{X} \in \{\mathrm{F}, \mathrm{G}^*, \mathrm{H}^*\}$.

4. *If* $r \geq 3$ *and* $c = r+2$*, then* $P = \mathrm{X} \xrightarrow{1} \mathrm{X}^*\left\{\xrightarrow{2} \mathrm{b}^*\right\}^{r-2} \xrightarrow{2} \mathrm{a}^* \xrightarrow{1} \mathrm{a}^*$*, where* $\mathrm{X} \in \{\mathrm{G}, \mathrm{H}\}$.

Proof. As in the proof of Theorem 3.6, see the dissertation of Nebelung [11]. □

3.7. Computing Artin patterns of p-groups

In both Algorithms 3.1 and 3.2, we made use of a subroutine `IsAdmissible()` which filters p-groups G with abelianization $G/G' \simeq (p,p)$ having a prescribed transfer kernel type (TKT). Since an algorithm of this kind is not implemented in MAGMA, we briefly communicate a succinct form of the code for this subroutine.

Algorithm 3.3. (Transfer kernel type.)

Input: a prime number p and a finite p-group G.

Code:

```
if ([ p,p] eq AbelianQuotientInvariants(G)) then
    x := G.1; y := G.2; // main generators
    A :=[ ] ; B :=[ ] ; // generators and transversal
    Append(~A,y);
    Append(~B,x);
    for e in[ 0..p-1] do
        Append(~A,x*y^e);
        Append(~B,y);
    end for;
    DG := DerivedSubgroup(G);
    nTotal := 0; nFixed := 0;
    TKT :=[ ] ;
    for i in[ 1..p+1] do
        M := sub<G|A[ i] ,DG>;
        DM := DerivedSubgroup(M);
        AQM,pr := M/DM;
        ImA := (A[ i] *B[ i] ^-1)^p*B[ i] ^p; // inner transfer
        ImB := B[ i] ^p; // outer transfer
        T := hom<G->AQM|<A[ i] ,(ImA)@pr>,<B[ i] ,(ImB)@pr>>;
```

```
        KT := sub<G|DG,Kernel(T)>;
        if KT eq G then // total kernel
           Append(~TKT,0);
           nTotal := nTotal+1;
        else
           for j in[1..p+1] do
              if A[j] in KT then
                 Append(~TKT,j);
                 if (i eq j) then // fixed point
                    nFixed := nFixed+1;
                 end if;
              end if;
           end for;
        end if;
     end for;
     image :=[];
     for i in[1..p+1] do
        if not (TKT[i] in image) then
           Append(~image,TKT[i]);
        end if;
     end for;
     occupation := #image;
     repetitions := 0; // maximal occupation number
     intersection := 0; // meet of repetitions and fixed points
     doublet := 0;
     for digit in[1..p+1] do
        counter := 0;
        for j in[1..#TKT] do
           if (digit eq TKT[j]) then
              counter := counter + 1;
           end if;
        end for;
        if (counter ge 2) then
           doublet := digit;
        end if;
        if (counter gt repetitions) then
           repetitions := counter;
        end if;
     end for;
     if (doublet ge 1) then
        if (doublet eq TKT[doublet]) then
           intersection := 1;
        end if;
     end if;
  end if;
```

Output: transfer kernel type `TKT`, number `nTotal` of total kernels, number `nFixed` of fixed points, and further invariants `occupation,repetitions,intersection` describing the orbit of the TKT.

The output of Algorithm 3.3 is used for the subroutine `IsAdmissible(G,p,t)` in dependence on the parameter flag t. When the root $R = \langle 243, 8\rangle$ is selected for the tree $\mathcal{T}(R)$ the return value is determined in the following manner

```
if (0 eq t) then
    return ((1 eq nTotal) and (2 eq nFixed)); // type c.21
elif (1 eq t) then
    return ((0 eq nTotal) and (3 eq nFixed)); // type E.8
elif (2 eq t) then
    return ((0 eq nTotal) and (2 eq nFixed) and (3 eq occupation)); // type E.9
end if;
```

For the root $R = \langle 243, 6\rangle$, we have

```
if (0 eq t) then
    return ((1 eq nTotal) and (0 eq nFixed)); // type c.18
elif (1 eq t) then
    return ((0 eq nTotal) and (1 eq nFixed)); // type E.6
elif (2 eq t) then
    return ((0 eq nTotal) and (0 eq nFixed) and (3 eq occupation)); // type E.14
end if;
```

3.8. Benefits and drawbacks of bottom up and top down techniques

In this chapter, we have presented several convenient ways of expressing information about *infinite sequences* of finite p-groups. Each of them has its benefits and drawbacks.

The *bottom up strategy* of constructing finite p-groups as successive extensions of a (metabelian or even abelian) starting group R, called the *root*, by recursive applications of the p-group algorithm by Newman [7] and O'Brien [8] has the benefit of visualizing the graph theoretic *root path* in the descendant tree $\mathcal{T}(R)$. Its implementation in MAGMA is incredibly stable and robust without surpassing any internal limits up to logarithmic orders of 95 and even more. Only the consumption of CPU time becomes considerable in such extreme regions.

The *top down strategy* of expressing finite p-groups as quotients of an infinite pro-p group with given pro-p presentation has the benefit of including nonmetabelian groups with arbitrary coclass $r \geq 3$, periodic mainline vertices in Algorithm 3.1, and sporadic Schur σ-leaves in Algorithm 3.2. The drawback is that the evaluation of the pro-p presentation in MAGMA exceeds the maximal permitted word length for nilpotency class $c \geq 36$.

Up to this point, we have not yet touched upon *parameterized polycyclic power-commutator presentations* [17]. For the root $R = \langle 243, 6\rangle$, the metabelian vertices G of the coclass tree $\mathcal{T}^{(2)}(R)$ with class $c = \mathrm{cl}(G) \geq 5$, down to depth $\mathrm{dp}(G) \leq 1$, can be presented in the form

$$\begin{aligned} G_c(\alpha,\beta) = \langle x, y, s_2, t_3, s_3, \ldots, s_c \mid \\ s_2 = [y,x],\ t_3 = [s_2, y],\ s_i = [s_{i-1}, x] \text{ for } 3 \leq i \leq c, \\ x^3 = s_c^{\alpha},\ y^3 s_3^{-2} s_4^{-1} = s_c^{\beta},\ s_i^3 = s_{i+2}^2 s_{i+3} \text{ for } 2 \leq i \leq c-3,\ s_{c-2}^3 = s_c^2 \ \rangle, \end{aligned} \tag{13}$$

where the parameters α and β depend on the transfer kernel type $\varkappa(G)$,

$$(\alpha,\beta) = \begin{cases} (0,0) & \text{for } \varkappa(G) \sim (0122),\ \text{c.18}, \\ (1,0) & \text{for } \varkappa(G) \sim (1122),\ \text{E.6}, \\ (0,1) \text{ or } (0,2) & \text{for } \varkappa(G) \sim (2122),\ \text{H.4}, \\ (1,1) \text{ or } (1,2) & \text{for } \varkappa(G) \sim (3122) \sim (4122),\ \text{E.14}. \end{cases} \tag{14}$$

This presentation has the benefit of including six periodic sequences with distinct transfer kernel types, and the drawback of being restricted to the fixed coclass 2.

4. The first 3-class towers of length 3

In our long desired disproof of the claim by Scholz and Taussky ([18], p. 41) concerning the 3-class tower of the imaginary quadratic field $F = \mathbb{Q}(\sqrt{-9748})$, we presented the first p-class towers with exactly three stages, for an odd prime p, in cooperation with Bush ([19], Cor. 4.1.1, p. 775). The underlying fields F were of type E.9 in its ground state, which admits two possibilities for the second 3-class group, $\mathfrak{M} \simeq \langle 2187, 302\rangle$ or $\langle 2187, 306\rangle$. Now we want to illustrate the way which led to the *fork topologies* in Theorem 3.5 by using the more convenient type E.8, where the group $\mathfrak{M}$ is unique for every state, in particular, $\mathfrak{M} \simeq \langle 2187, 304\rangle$ for the ground state.

Remark 4.1. Concerning the notation, we are going to use *logarithmic type invariants* of abelian 3-groups, for instance $(21) \hat{=} (9,3)$, $(32) \hat{=} (27,9)$, $(43) \hat{=} (81,27)$, and $(54) \hat{=} (243,81)$.

Let $F = \mathbb{Q}(\sqrt{d})$ be an *imaginary* quadratic number field with 3-class group $\mathrm{Cl}_3 F \simeq (3,3)$, and let $E_1, \ldots, E_4$ be the unramified cyclic cubic extensions of F.

Theorem 4.1. *(First towers of type* E.8.*) Let the capitulation of 3-classes of F in $E_1, \ldots, E_4$ be of type $\varkappa_1 F \sim (\mathbf{1}, \mathbf{2}, \mathbf{3}, 1)$, which is called type* E.8. *Assume further that the 3-class groups of $E_1, \ldots, E_4$ possess the abelian type invariants $\tau_1 F \sim [T_1, 21, 21, 21]$, where $T_1 \in \{32, 43, 54\}$.*

Then, the length of the 3-class tower of F is precisely $\ell_3 F = 3$.

Proof. We employ the p-group generation algorithm [7, 8] for searching the Artin pattern $\mathrm{AP}(F) = (\tau_1 F, \varkappa_1 F)$ among the descendants of the root $R := C_3 \times C_3 = \langle 9, 2\rangle$ in the tree $\mathcal{T}(R)$. After two steps, $\langle 9,2\rangle \leftarrow \langle 27,3\rangle \leftarrow \langle 243,8\rangle$, we find the next root $U_5 := \langle 243, 8\rangle$ of the unique relevant coclass tree $\mathcal{T}^{(2)}(U_5)$, using the assigned simple TKT E.8, $\varkappa_3 = (1231)$, and its associated scaffold TKT c.21, $\varkappa_0 = (0231)$. Finally, the first component $T_1 = \tau_1(1) \in \{32, 43, 54\}$ of the

TTT provides the break-off condition, according to ([13], Thm. 1.21, p. 79), respectively, Theorem M in ([20], p. 14), and we get $\mathfrak{M} \simeq \langle 2187, 304\rangle = \langle 729, 54\rangle - \#1;4$ for the ground state $T_1 = (32)$, $\mathfrak{M} \simeq \langle 6561, 2050\rangle - \#1;2$ for the first excited state $T_1 = (43)$, and $\mathfrak{M} \simeq \langle 6561, 2050\rangle (-\#1;1)^2 - \#1;2$ for the second excited state $T_1 = (54)$, where $\langle 2187, 303\rangle = \langle 729, 54\rangle - \#1;3$ and $\langle 6561, 2050\rangle = \langle 2187, 303\rangle - \#1;1$. The situation is visualized by **Figure 2**, where the three metabelianizations $\mathfrak{M} \simeq G/G''$ of the 3-tower group G, for the ground state and two excited states, are emphasized with red color. **Figure 2**, showing the second 3-class groups $\mathfrak{M}$, was essentially known to Ascione in 1979 [21, 22], and to Nebelung in 1989 [11]. Compare the historical remarks ([2], section 3, p. 163).

In the next three **Figures 5–7**, which were unknown until 2012, we present the decisive breakthrough establishing the first rigorous proof for three-stage towers of 3-class fields. The key ingredient is the discovery of periodic bifurcations ([2], section 3, p. 163) in the complete descendant tree $\mathcal{T}(U_5)$ which is of considerably higher complexity than the coclass tree $\mathcal{T}^{(2)}(U_5)$.

For the ground state $T_1 = (32)$, the first bifurcation yields the cover

$$\mathrm{cov}(\mathfrak{M}) = \{\mathfrak{M}, \langle 6561, 622\rangle\}$$

of $\mathfrak{M} \simeq \langle 2187, 304\rangle$. The relation rank $d_2\mathfrak{M} = 3$ eliminates $\mathfrak{M}$ as a candidate for the 3-tower group G, according to the Corollary ([20], p. 7) of the Shafarevich Theorem ([13], Thm. 1.3, pp. 75–76), and we end up getting $G \simeq \langle 6561, 622\rangle = \langle 729, 54\rangle - \#2;4$ with a siblings topology

$$\mathrm{E} \stackrel{1}{\rightarrow} \mathrm{c} \stackrel{2}{\leftarrow} \mathrm{E}$$

which describes the relative location of $\mathfrak{M}$ and G.

For the first excited state $T_1 = (43)$, the second bifurcation yields the cover

$$\mathrm{cov}(\mathfrak{M}) = \{\mathfrak{M}, \langle 6561, 621\rangle - \#1;1 - \#1;2, \langle 6561, 621\rangle - \#1;1 - \#2;2\}$$

of $\mathfrak{M} \simeq \langle 6561, 2050\rangle - \#1;2$, where $\langle 6561, 621\rangle = \langle 729, 54\rangle - \#2;3$. The relation rank $d_2 = 3$ eliminates $\mathfrak{M}$ and $\langle 6561, 621\rangle - \#1;1 - \#1;2$ as candidates for the 3-tower group G, according to Shafarevich, and we get the unique $G \simeq \langle 6561, 621\rangle - \#1;1 - \#2;2$ with fork topology

$$\mathrm{E} \stackrel{1}{\rightarrow} \left\{\mathrm{c} \stackrel{1}{\rightarrow}\right\}^2 \mathrm{c} \left\{\stackrel{2}{\leftarrow} \mathrm{c} \stackrel{1}{\leftarrow} \mathrm{c}\right\} \stackrel{2}{\leftarrow} \mathrm{E}.$$

Similarly, the second excited state $T_1 = (54)$ yields a more complex advanced fork topology

$$\mathrm{E} \stackrel{1}{\rightarrow} \left\{\mathrm{c} \stackrel{1}{\rightarrow}\right\}^4 \mathrm{c} \left\{\stackrel{2}{\leftarrow} \mathrm{c} \stackrel{1}{\leftarrow} \mathrm{c}\right\}^2 \stackrel{2}{\leftarrow} \mathrm{E}.$$

Figure 5 impressively shows that entering the unnoticed secret door, which is provided by the bifurcation at the vertex $\langle 729, 54\rangle$, immediately leads to the long desired 3-tower group $G \simeq \langle 6561, 622\rangle = \langle 729, 54\rangle - \#2;4$ of the imaginary quadratic field $F = \mathbb{Q}(\sqrt{-34\,867})$. The siblings topology is emphasized with red color, and the projection $G \rightarrow \mathfrak{M} \simeq G/G''$ is drawn in blue color.

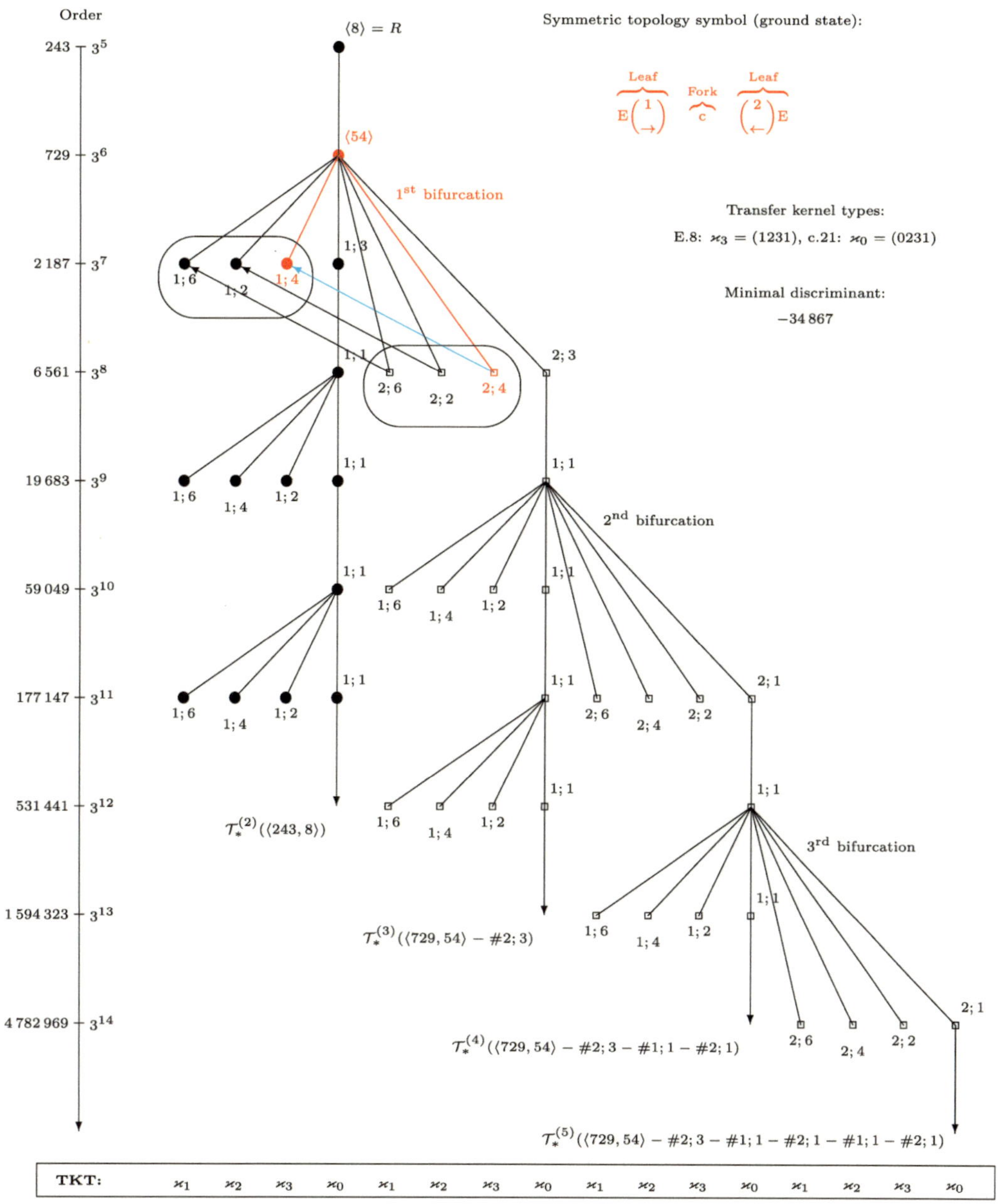

Figure 5. Tree topology of type E in the ground state.

In **Figure 6**, we see that the path to the 3-tower group $G \simeq \langle 729, 54\rangle - \#2;3 - \#1;1 - \#2;2$ of the imaginary quadratic field $F = \mathbb{Q}(\sqrt{-370\,740})$ contains two bifurcations at $\langle 729, 54\rangle$ and $\langle 729, 54\rangle - \#2;3 - \#1;1$. As before, the fork topology is emphasized with red color, and the projection $G \to \mathfrak{M} \simeq G/G''$ is drawn in blue color. Two projection arrows of type E.9 are black.

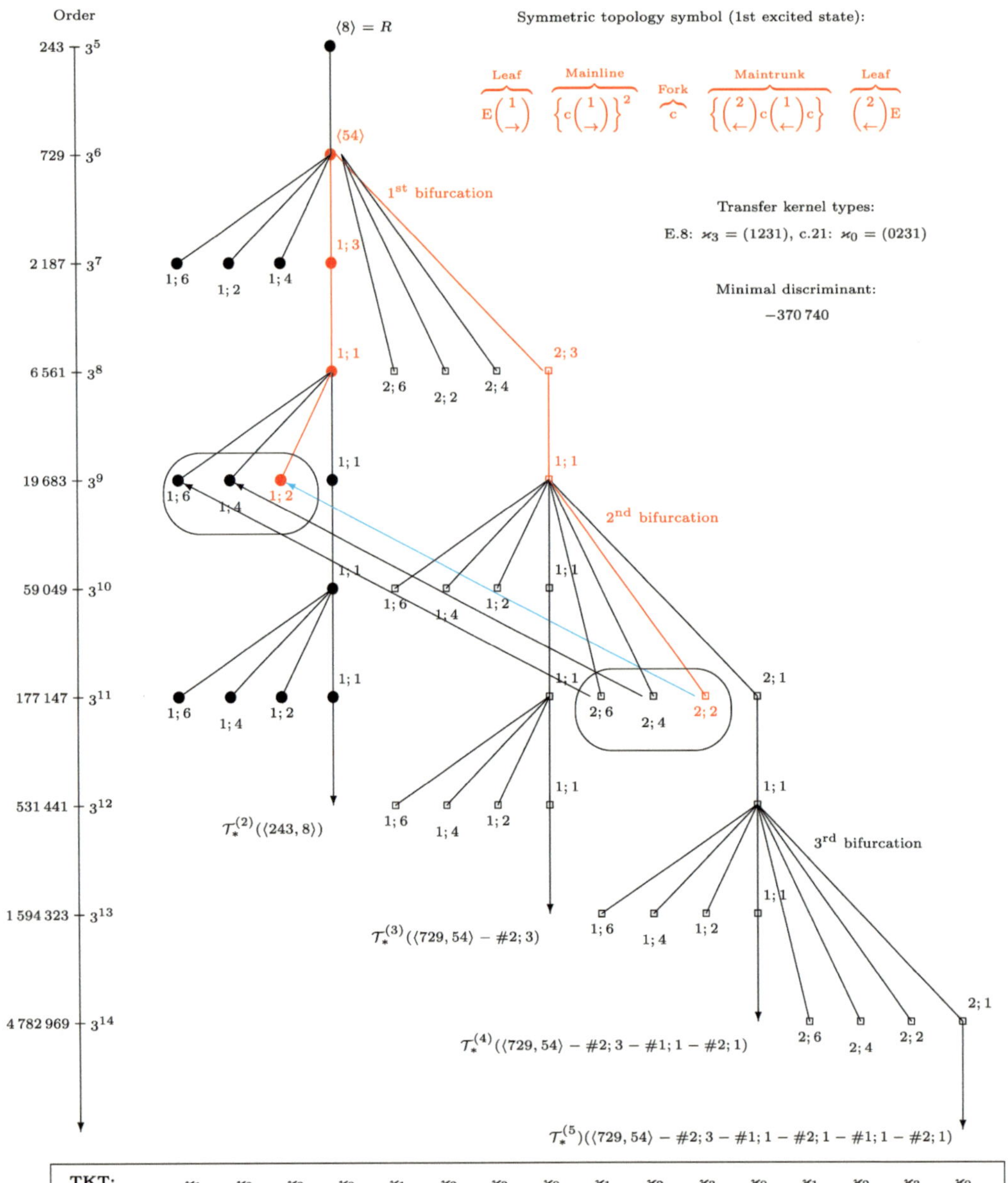

Figure 6. Tree topology of type E in the first excited state.

Figure 7 shows the path to the 3-tower group $G \simeq \langle 729, 54\rangle - \#2;3 - \#1;1 - \#2;1 - \#1;1 - \#2;2$ of the imaginary quadratic field $F = \mathbb{Q}(\sqrt{-4\,087\,295})$. It requires three bifurcations at $\langle 729, 54\rangle$, $\langle 729, 54\rangle - \#2;3 - \#1;1$, and $\langle 729, 54\rangle - \#2;3 - \#1;1 - \#2;1 - \#1;1$. Again, the fork topology is emphasized with red color, and the projection $G \to \mathfrak{M} \simeq G/G''$ is drawn in blue color.

http://dx.doi.org/10.5772/intechopen.68703

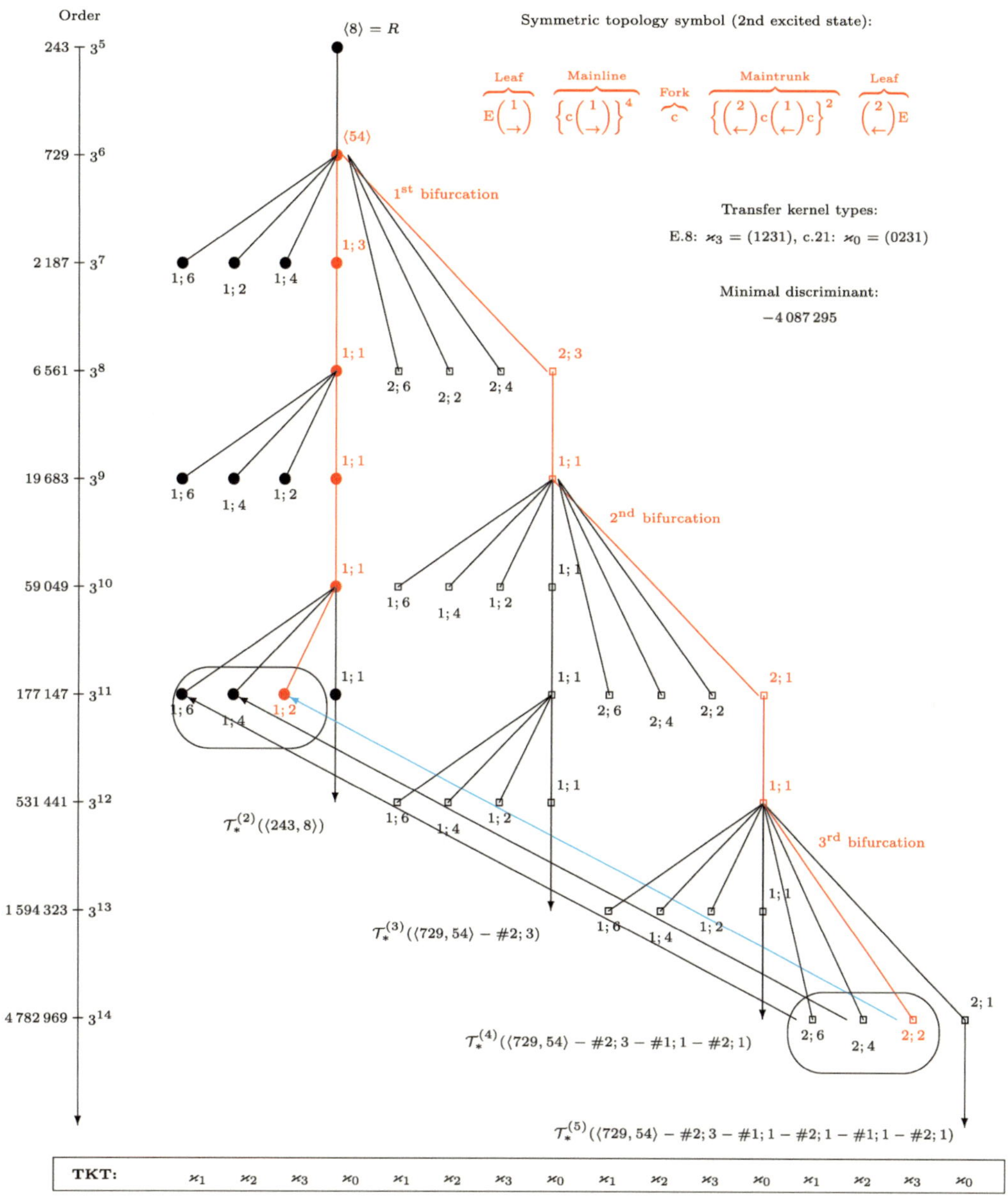

Figure 7. Tree topology of type E in the second excited state.

5. Future developments

Fork topologies with significantly higher complexity and step sizes up to 3 and even 4 will be investigated in cooperation with M. F. Newman [23] for finite 3-groups with TKT F.

Acknowledgements

The author gratefully acknowledges support from the Austrian Science Fund (FWF): P 26008-N25. Indebtedness is expressed to M. F. Newman from the Australian National University, Canberra, for valuable help in providing evidence of Theorems 3.2, 3.3 and 3.5.

Author details

Daniel C. Mayer

Address all correspondence to: algebraic.number.theory@algebra.at

Austrian Science Fund (FWF), Graz, Austria

References

[1] Gamble G, Nickel W, O'Brien EA. ANU p-Quotient — p-Quotient and p-Group Generation Algorithms. 2006. An accepted GAP package, available also in MAGMA

[2] Mayer DC. Periodic bifurcations in descendant trees of finite p-groups. Advances in Pure Mathematics. 2015;**5**(4):162–195. DOI: 10.4236/apm.2015.54020. Special Issue on Group Theory, March 2015

[3] Besche HU, Eick B, O'Brien EA. The SmallGroups Library—a Library of Groups of Small Order. 2005. An accepted and refereed GAP package, available also in MAGMA

[4] du Sautoy M. Counting p-groups and nilpotent groups. Publications Mathématiques de l'Institut des Hautes Études Scientifiques. 2000;**92**:63–112

[5] Eick B, Leedham-Green C. On the classification of prime-power groups by coclass. Bulletin of the London Mathematical Society. 2008;**40**(2):274–288

[6] Mayer DC. New number fields with known p-class tower. Tatra Mountains Mathematical Publications. 2015;**64**:21–57. DOI: 10.1515/tmmp-2015-0040. Special Issue on Number Theory and Cryptology 15

[7] Newman MF. Determination of groups of prime-power order. In: Group Theory, Canberra, 1975, Lecture Notes in Math. Vol. 573. Berlin: Springer; 1977. pp. 73–84

[8] O'Brien EA. The p-group generation algorithm. Journal of Symbolic Computation. 1990;**9**: 677–698

[9] Mayer DC. Artin transfer patterns on descendant trees of finite p-groups. Advances in Pure Mathematics. 2016;**6**(2):66–104. DOI: 10.4236/apm.2016.62008. Special Issue on Group Theory Research, January 2016.

[10] The MAGMA Group. MAGMA Computational Algebra System, Version 2.22-7, Sydney. 2016. (Available from: http://magma.maths.usyd.edu.au)

[11] Nebelung B. Klassifikation metabelscher 3-Gruppen mit Faktorkommutatorgruppe vom Typ (3, 3) und Anwendung auf das Kapitulationsproblem [Inauguraldissertation]. Universität zu Köln; 1989

[12] Mayer DC, Newman MF. Finite 3-groups as Viewed from Class Field theory. Scotland, UK: Groups St. Andrews, Univ. of St. Andrews, Fife; 2013. Contributed presentation delivered on 11 August 2013. http://www.algebra.at/GroupsStAndrews2013.pdf

[13] Mayer DC. Recent progress in determining p-class field towers. Gulf Journal of Mathematics (Dubai, UAE). 2016;**4**(4):74–102. ISSN 2309-4966

[14] Shafarevich IR. Extensions with prescribed ramification points (Russian). Publications Mathématiques de l'Institut des Hautes Études Scientifiques.1964;**18**:71–95. (English transl. by J. W. S. Cassels in Amer. Math. Soc. Transl., II. Ser., 59 (1966), 128–149.)

[15] Mayer DC. Periodic sequences of p-class tower groups. Journal of Applied Mathematics and Physics. 2015;**3**(7):746–756. DOI: 10.4236/jamp.2015.37090. First International Conference on Groups and Algebras, 2015, Shanghai.

[16] Mayer DC. Annihilator ideals of two-generated metabelian p-groups. Journal of Algebra and its Applications (JAA). (arXiv: 1603.09288v1 [math.GR] 30 Mar 2016.)

[17] Mayer DC. Power-commutator presentations for infinite sequences of 3-groups, preprint. 2013. (Available from: https://www.researchgate.net/publication/256459683_Power-commutator_presentations_for_infinite_sequences_of_3-groups)

[18] Scholz A, Taussky O. Die Hauptideale der kubischen Klassenkörper imaginär quadratischer Zahlkörper: ihre rechnerische Bestimmung und ihr Einfluß auf den Klassenkörperturm. J. Reine Angew. Math. 1934;**171**:19–41

[19] Bush MR, Mayer DC. 3-class field towers of exact length 3. Journal of Number Theory. 2015;**147**:766–777. DOI: 10.1016/j.jnt.2014.08.010

[20] Mayer DC. Recent progress in determining p-class field towers, 1st International Colloquium of Algebra, Number Theory, Cryptography and Information Security 2016, Faculté Polydisciplinaire de Taza, Université Sidi Mohamed Ben Abdellah, Fès, Morocco, invited keynote delivered on 12 November 2016. Available from: http://www.algebra.at/ANCI2016DCM.pdf

[21] Ascione JA. On 3-groups of second maximal class [Ph.D. Thesis]. Canberra: Austral. National Univ.; 1979

[22] Ascione JA. On 3-groups of second maximal class. Bulletin of the Australian Mathematical Society. 1980;**21**:473–474

[23] Mayer DC, Newman MF. Finite 3-groups with transfer kernel type F, in preparation. Available from: http://www.algebra.at/TransferSectionF.pdf

Chapter 6

Monophonic Distance in Graphs

P. Titus and A.P. Santhakumaran

Additional information is available at the end of the chapter

http://dx.doi.org/10.5772/intechopen.68668

Abstract

For any two vertices u and v in a connected graph G, a $u - v$ path is a *monophonic path* if it contains no chords, and the *monophonic distance* $d_m(u, v)$ is the length of a longest $u - v$ monophonic path in G. For any vertex v in G, the *monophonic eccentricity* of v is $e_m(v) = max\ \{d_m(u, v) : u \in V\}$. The subgraph induced by the vertices of G having minimum monophonic eccentricity is the *monophonic center* of G, and it is proved that every graph is the monophonic center of some graph. Also it is proved that the monophonic center of every connected graph G lies in some block of G. With regard to convexity, this monophonic distance is the basis of some detour monophonic parameters such as detour monophonic number, upper detour monophonic number, forcing detour monophonic number, etc. The concept of detour monophonic sets and detour monophonic numbers by fixing a vertex of a graph would be introduced and discussed. Various interesting results based on these parameters are also discussed in this chapter.

Keywords: monophonic path, monophonic distance, detour monophonic number, upper detour monophonic number, forcing detour monophonic number, vertex detour monophonic number, upper vertex detour monophonic number, forcing vertex detour monophonic number

1. Introduction

In this chapter, we consider a finite connected graph $G = (V(G), E(G))$ having no loops and multiple edges. The *order* and *size* of G are denoted by p and q, respectively. Distance in graphs is a wide branch of graph theory having numerous scientific and real-life applications. There are many kinds of distances in graphs found in literature. For any two vertices u and v in G, the *distance* $d(u, v)$ from u to v is defined as the length of a shortest $u - v$ path in G. The *eccentricity* $e(v)$ of a vertex v in G is the maximum distance from v to a vertex of G. The *radius rad* G of G is the minimum eccentricity among the vertices of G, while the *diameter diam* G of G is the maximum eccentricity among the vertices of G. The distance between two vertices is a fundamental concept in pure graph theory, and this distance is a metric on the vertex set of G. More results related to this distance are found in Refs. [1–9]. This distance is used to study the central concepts like center, median, and centroid of a graph [10–22]. With regard to convexity, this distance is the basis of some geodetic parameters such as geodetic number, connected geodetic number, upper geodetic number and forcing geodetic number [23–32]. The geodesic graphs, extremal graphs, distance regular graphs and distance transitive graphs are some important classes based on the distance in graphs [33, 34]. These concepts have interesting applications in location theory and convexity theory. The *neighborhood* of a vertex v is the set $N(v)$ consisting of all vertices u which are adjacent with v. A vertex v is an *extreme vertex* if the subgraph induced by its neighbors is complete.

The detour distance, which is defined to be the length of a longest path between two vertices of a graph, is also a metric on the vertex set of G [35, 36]. For any two vertices u and v in G, the *detour distance* $D(u, v)$ from u to v is defined as the length of a longest $u - v$ path in G. The *detour eccentricity* $e_D(v)$ of a vertex v in G is the maximum detour distance from v to a vertex of G. The *detour radius* $rad_D\, G$ of G is the minimum detour eccentricity among the vertices of G, while the *detour diameter* $diam_D\, G$ of G is the maximum detour eccentricity among the vertices of G. With regard to detour convexity, the detour number of a graph was introduced and studied in Refs. [25, 37]. The detour concepts and colorings are widely used in the channel assignment problem in FM radio technology and also in certain molecular problems in theoretical chemistry.

The parameter geodetic (detour) number of a graph is global in the sense that there is exactly one geodetic (detour) number for a graph. The concept of geodetic (detour) sets and geodetic (detour) numbers by fixing a vertex of a graph was also introduced and discussed in Refs. [38–42]. With respect to each vertex of a graph, there is a geodetic (detour) number, and so there will be at most as many geodetic (detour) numbers as there are vertices in the graph.

2. Monophonic distance

Definition 2.1. *A chord of a path* $u_1, u_2, \ldots, u_n$ *in a connected graph* G *is an edge* $u_i u_j$ *with* $j \geq i + 2$. *A* $u - v$ *path* P *is called a monophonic path if it is a chordless path. The length of a longest* $u - v$ *monophonic path is called the monophonic distance from* u *to* v, *and it is denoted by* $d_m(u, v)$. *A* $u - v$ *monophonic path with its length equal to* $d_m(u, v)$ *is known as a* $u - v$ *monophonic.*

Example 2.2. Consider the graph G given in **Figure 1**. It is easily verified that $d(v_1, v_4) = 2$, $D(v_1, v_4) = 6$, and $d_m(v_1, v_4) = 4$. Thus the monophonic distance is different from both the distance and the detour distance. The monophonic path $P : v_1, v_2, v_8, v_7, v_4$ is $v_1 - v_4$ monophonic while the monophonic path $Q : v_1, v_3, v_4$ is not $v_1 - v_4$ monophonic.

The usual distance d and the detour distance D are metrics on the vertex set V of a connected graph G, whereas the monophonic distance d_m is not a metric on V. For the graph G given in **Figure 1**, $d_m(v_4, v_6) = 5$, $d_m(v_4, v_5) = 1$ and $d_m(v_5, v_6) = 1$. Hence $d_m(v_4, v_6) > d_m(v_4, v_5) + d_m(v_5, v_6)$, and so the triangle inequality is not satisfied.

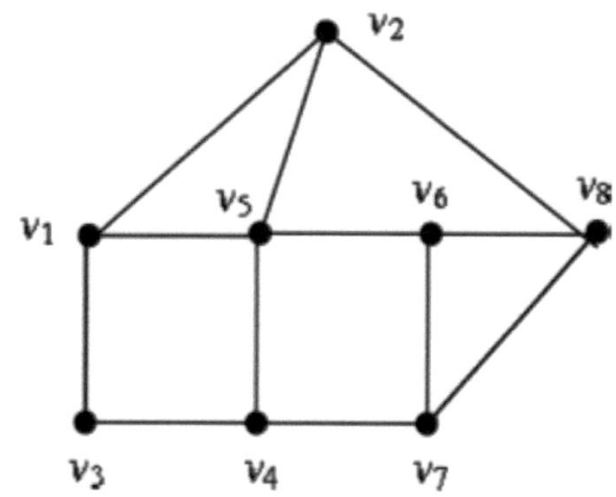

Figure 1. The graph G in Example 2.2.

The following result is an easy consequence of the respective definitions.

Proposition 2.3. *Let u and v be any two vertices in a graph G of order p. Then*

$$0 \le d(u, v) \le d_m(u, v) \le D(u, v) \le p - 1.$$

Result 2.4. Let u and v be any two vertices in a connected graph G. Then

(i) $d_m(u, v) = 0$ if and only if $u = v$.

(ii) $d_m(u, v) = 1$ if and only if uv is an edge of G.

(iii) $d_m(u, v) = p - 1$ if and only if G is the path with endvertices u and v.

(iv) $d(u, v) = d_m(u, v) = D(u, v)$ if and only if G is a tree.

Definition 2.5. *For any vertex v in a connected graph G, the monophonic eccentricity of v is* $e_m(v) = max \{d_m(u, v) : u \in V\}$. *A vertex u of G such that* $d_m(u, v) = e_m(v)$ *is called a monophonic eccentric vertex of v. The monophonic radius and monophonic diameter of G are defined by* $rad_m\, G = min \{e_m(v) : v \in V\}$ *and* $diam_m\, G = max \{e_m(v) : v \in V\}$, *respectively.*

Example 2.6. Table 1 shows the monophonic distance between the vertices and also the monophonic eccentricities of vertices of the graph G given in **Figure 1**. It is to be noted that $rad_m\, G = 3$ and $diam_m\, G = 5$.

Remark 2.7. In any connected graph, the eccentricity of every two adjacent vertices differs by at most 1. However, this is not true in the case of monophonic distance. For the graph G given in **Figure 1**, $e_m(v_5) = 3$ and $e_m(v_6) = 5$.

Note 2.8. Any two vertices u and v in a tree T are connected by a unique path, and so $d(u, v) = d_m(u, v) = D(u, v)$. Hence $rad\, T = rad_m\, T = rad_D\, T$ and $diam\, T = diam_m\, T = diam_D\, T$. The monophonic radius and the monophonic diameter of some standard graphs are given in **Table 2**.

$d_m(v_i, v_j)$	v_1	v_2	v_3	v_4	v_5	v_6	v_7	v_8	$e_m(v)$
v_1	0	1	1	4	1	4	3	4	4
v_2	1	0	4	3	1	5	4	1	5
v_3	1	4	0	1	2	4	4	4	4
v_4	4	3	1	0	1	5	1	4	5
v_5	1	1	2	1	0	1	3	3	3
v_6	4	5	4	5	1	0	1	1	5
v_7	3	4	4	1	3	1	0	1	4
v_8	4	1	4	4	3	1	1	0	4

Table 1. Monophonic eccentricities of the vertices of the graph G in **Figure 1**.

Graph G	K_p	C_p	$W_{1,p-1}$ $(p \geq 4)$	$K_{1,p-1}$ $(p \geq 2)$	$K_{m,n}$ $(m,n \geq 2)$	P_p
$rad_m\, G$	1	$p-2$	1	1	2	$\left\lfloor \frac{p}{2} \right\rfloor$
$diam_m\, G$	1	$p-2$	$p-3$	2	2	$p-1$

Table 2. Monophonic radius and monophonic diameter of some standard graphs.

The next theorem follows from Proposition 2.3.

Theorem 2.9. *For a connected graph G, the following results hold:*

a. $e(v) \leq e_m(v) \leq e_D(v)$ *for any vertex* v *in* G.

b. $rad\, G \leq rad_m\, G \leq rad_D\, G$.

c. $diam\, G \leq diam_m\, G \leq diam_D\, G$.

Theorem 2.10. *(a) If a, b and c are integers with* $3 \leq a \leq b \leq c$, *then there exists a connected graph* G *such that* $rad\, G = a$, $rad_m\, G = b$ *and* $rad_D\, G = c$.

(b) If a, b and c are integers with $5 \leq a \leq b \leq c$, *then there exists a connected graph* G *such that* $diam\, G = a$, $diam_m\, G = b$ *and* $diam_D\, G = c$.

Proof. (a) The result is proved by considering three cases.

Case (i) $3 \leq a = b = c$. Consider $G = P_{2a+1}$, the path of order $2a + 1$. It is clear that $rad\, G = rad_m\, G = rad_D\, G = a$.

Case (ii) $3 \le a \le b < c$. Let $F_1 : u_1, u_2,\ldots, u_{a-1}$ and $F_2 : v_1, v_2,\ldots, v_{a-1}$ be two copies of the path P_{a-1} of order $a - 1$. Let $F_3 : w_1, w_2,\ldots, w_{b-a+3}$ and $F_4 : z_1, z_2,\ldots, z_{b-a+3}$ be two copies of the path P_{b-a+3} of order $b-a+3$, and $F_5 = K_{c-b+1}$ the complete graph of order $c - b + 1$ with $V(F_5) = \{x_1, x_2,\ldots, x_{c-b+1}\}$. We construct the graph G as follows: (i) identify the vertices x_1 in F_5 and w_1 in F_3; also identify the vertices x_{c-b+1} in F_5 and z_1 in F_4; (ii) identify the vertices w_{b-a+3} in F_3 and u_2 in F_1, and identify the vertices z_{b-a+3} in F_4 and v_2 in F_2; and (iii) join each vertex w_i $(1 \le i \le b-a+2)$ in F_3 and u_1 in F_1, and join each vertex z_i $(1 \le i \le b-a+2)$ in F_4 and v_1 in F_2. The resulting graph G is shown in **Figure 2**. It is easily verified that $e(v) = a$ if $v \in V(F_5)$; $e(v) > a$ if $v \in V(G) - V(F_5)$, $e_m(v) = b$ if $v \in V(F_5)$; $e_m(v) > b$ if $v \in V(G) - V(F_5)$ and $e_D(v) = c$ if $v \in V(F_5)$; and $e_D(v) > c$ if $v \in V(G) - V(F_5)$. It follows that $rad\ G = a$, $rad_m\ G = b$, and $rad_D\ G = c$.

Case (iii) $3 \le a < b = c$. Let $E_1 : v_1, v_2,\ldots, v_{2a+1}$ be a path of order $2a + 1$. Let $E_2 : u_1, u_2,\ldots, u_{b-a+3}$ and $E_3: w_1, w_2,\ldots, w_{b-a+3}$ be two copies of the path P_{b-a+3} of order $b - a + 3$, and let E_i $(4 \le i \le 2(b - a) + 3)$ be $2(b - a)$ copies of K_1. We construct the graph G as follows: (i) identify the vertices v_{a+1} in E_1, u_1 in E_2, and w_1 in E_3; (ii) identify the vertices v_{a-1} in E_1 and u_{b-a+3} in E_2, and identify the vertices v_{a+3} in E_1 and w_{b-a+3} in E_3; and (iii) join each E_i $(4 \le i \le b - a + 3)$ with v_{a+1} in E_1 and u_{i-1} in E_2, and join each E_i $(b-a+4 \le i \le 2(b-a)+3)$ with v_{a+1} in E_1 and $w_{i-b+a-1}$ in E_3. The resulting graph G is shown in **Figure 3**.

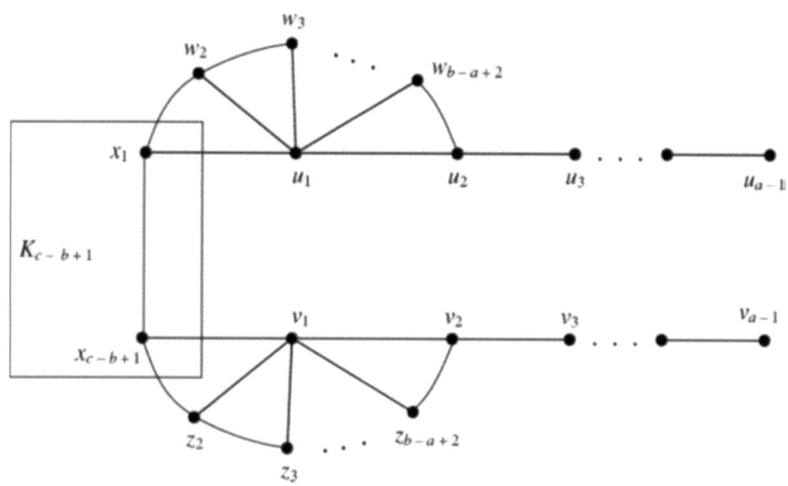

Figure 2. A graph G in Case (ii) of Theorem 2.10(a).

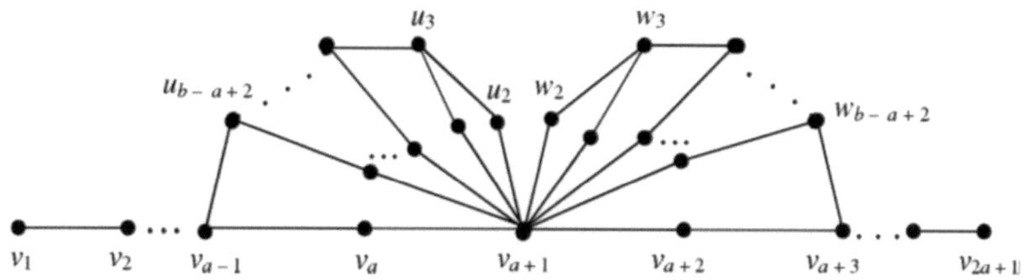

Figure 3. A graph G in Case (iii) of Theorem 2.10(a).

It is easily verified that $e(v_{a+1}) = a$; $e(v) > a$ if $v \in V(G) - \{v_{a+1}\}$; $e_m(v_{a+1}) = b$; $e_m(v) > b$ if $v \in V(G) - \{v_{a+1}\}$, and $e_D(v_{a+1}) = c$; and $e_D(v) > c$ if $v \in V(G) - \{v_{a+1}\}$. It follows that $rad\ G = a$, $rad_m\ G = b$, and $rad_D\ G = c$.

(b) This result is also proved by considering three cases.

Case (i) $5 \le a = b = c$. Let G be a path of order a+1. Then $diam\ G = diam_m\ G = diam_D\ G = a$.

Case (ii) $5 \le a \le b < c$. Let $F_1 : u_1, u_2, \ldots, u_{a-1}$ be the path P_{a-1} of order $a - 1$; $F_2 : w_1, w_2, \ldots, w_{b-a+3}$ be the path P_{b-a+3} of order $b - a + 3$; and $F_3 = K_{c-b+1}$ be the complete graph of order $c - b + 1$ with $V(F_3) = \{x_1, x_2, \ldots, x_{c-b+1}\}$. We construct the graph G as follows: (i) identify the vertices x_1 in F_3 and w_1 in F_2, and identify the vertices w_{b-a+3} in F_2 and u_2 in F_1, and (ii) join each vertex w_i $(1 \le i \le b - a + 2)$ in F_2 and u_1 in F_1. The resulting graph G is shown in **Figure 4**. It is easily verified that $e(v) = a$ if $v \in (V(F_3) - \{x_1\}) \cup \{u_{a-1}\}$; $e(v) < a$ if $v \in V(F_2) \cup (V(F_1) - \{u_{a-1}\})$, and $e_m(v) = b$ if $v \in (V(F_3) - \{x_1\}) \cup \{u_{a-1}\}$; $e_m(v) < b$ if $v \in V(F_2) \cup (V(F_1) - \{u_{a-1}\})$, and $e_D(v) = c$ if $v \in (V(F_3) - \{x_1\}) \cup \{u_{a-1}\}$; and $e_D(v) < c$ if $v \in V(F_2) \cup (V(F_1) - \{u_{a-1}\})$. It follows that $diam\ G = a$, $diam_m\ G = b$ and $diam_D\ G = c$.

Case (iii) $5 \le a < b = c$. Let $E_1 : v_1, v_2, \ldots, v_{a+1}$ be a path of order $a + 1$; $E_2 : w_1, w_2, \ldots, w_{b-a+3}$ be another path of order $b - a + 3$; and E_i $(3 \le i \le b - a + 2)$ be $b - a$ copies of K_1. Let G be the graph obtained from E_i for $i = 1, 2, \ldots, b - a + 2$ by identifying the vertices v_{a-2} and v_a of E_1 with w_1 and w_{b-a+3} of E_2, respectively, and joining each E_i $(3 \le i \le b - a + 2)$ with v_{a-2} and w_i. The graph G is shown in **Figure 5**.

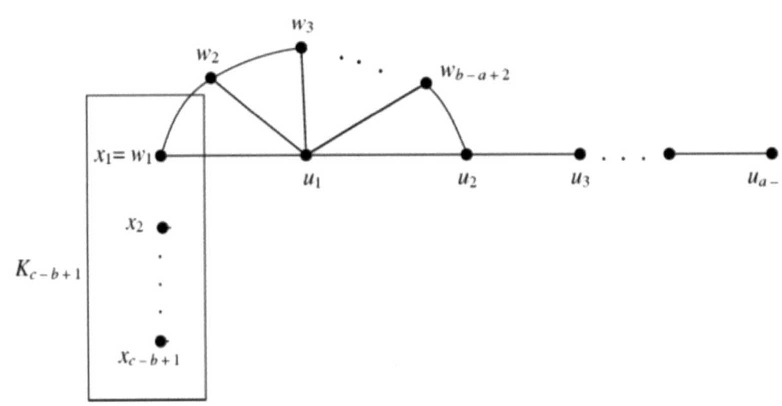

Figure 4. A graph G in Case (ii) of Theorem 2.10(b).

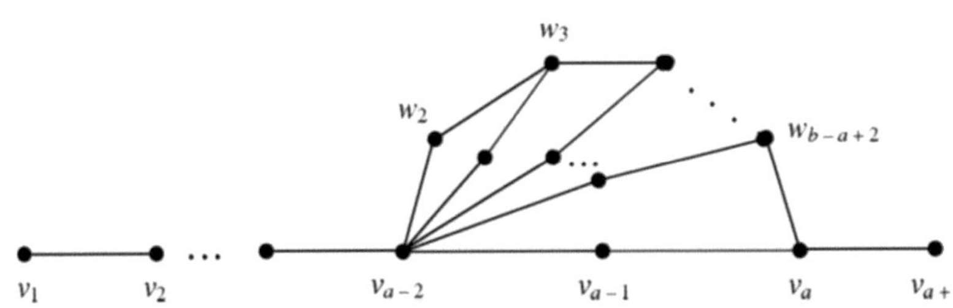

Figure 5. A graph G in Case (iii) of Theorem 2.10(b).

It is easily verified that $e(v) = a$ if $v \in \{v_1, v_{a+1}\}$; $e(v) \le a$ if $v \in V(G) - \{v_1, v_{a+1}\}$, and $e_m(v) = b$ if $v \in \{v_1, v_{a+1}\}$; $e_m(v) \le b$ if $v \in V(G) - \{v_1, v_{a+1}\}$, and $e_D(v) = c$ if $v \in \{v_1, v_{a+1}\}$; and $e_D(v) \le c$ if $v \in V(G) - \{v_1, v_{a+1}\}$. It follows that $rad\ G = a$, $rad_m\ G = b$ and $rad_D\ G = c$.

For any connected graph G, the inequalities $rad\ G \le diam\ G \le 2\ rad\ G$ and $rad_D\ G \le diam_D\ G \le 2\ rad_D\ G$ hold. However, this is not true in the case of monophonic radius and monophonic diameter. For example, when the graph G is the wheel $W_{1,p-1}$ $(p \ge 6)$, it is easily seen that $rad_m\ G = 1$ and $diam_m\ G = p - 3 \ge 3$ so that $diam_m\ G > 2\ rad_m\ G$.

It is proved in Ref. [6] that if a and b are any two positive integers such that $a \le b \le 2a$, then there is a connected graph G with $rad\ G = a$ and $diam\ G = b$. Also, it is proved in Ref. [35] that if a and b are any two positive integers such that $a \le b \le 2a$, then there is a connected graph G with $rad_D\ G = a$ and $diam_D\ G = b$.

Now, the following theorem gives a realization result for $rad_m\ G \le diam_m\ G$.

Theorem 2.11. *If a and b are positive integers with $a \le b$, then there exists a connected graph G such that $rad_m\ G = a$ and $diam_m\ G = b$.*

Proof. This result is proved by considering three cases.

Case (i) $a = b \ge 1$. Let G be the cycle C_{a+2}. Then $rad_m\ G = a$ and $diam_m\ G = b$.

Case (ii) $a < b \le 2a$. Let $C_1 : u_1, u_2, \ldots, u_{a+2}, u_1$ be a cycle of order $a + 2$ and $C_2 : v_1, v_2, \ldots, v_{b-a+2}, v_1$ be a cycle of order $b - a + 2$. Let G be the graph obtained by identifying the vertex u_1 of C_1 and v_1 of C_2. Since $b \le 2a$, it follows that $b - a + 2 \le a + 2$. It is clear that $d_m(u_1, x) \le a$ for any x in G and $d_m(u_1, u_{a+1}) = a$. Therefore, $e_m(u_1) = a$. Also, it is clear that there is no vertex x with $e_m(x) < a$ and so $rad_m\ G = a$. It is clear that $d_m(u_3, v_3) = b$ and $d_m(u_3, x) \le b$ for any vertex x in G and so $e_m(u_3) = b$. Also, it is easy to see that $e_m(x) \le b$ for every vertex x in G so that $diam_m\ G = b$.

Case (iii) $b > 2a$. Let G be the graph obtained by identifying the central vertex of the wheel $W = K_1 + C_{b+2}$ $(b \ge 2)$ and an endvertex of the path P_{2a}. Since $b > 2a$, $e_m(x) = b$ for any vertex $x \in V(C_{b+2})$. Also, $a \le e_m(x) \le 2a$ for any vertex $x \in V(P_{2a})$ and $e_m(v_a) = a$. Hence $rad_m\ G = a$ and $diam_m\ G = b$.

2.1. Monophonic center and monophonic periphery

Definition 2.12. *A vertex v in a connected graph G is called a monophonic central vertex if $e_m(v) = rad_m\ G$, and the subgraph induced by the monophonic central vertices of G is the monophonic center $C_m(G)$ of G. A vertex v in G is called a monophonic peripheral vertex if $e_m(v) = diam_m\ G$, and the subgraph induced by the monophonic peripheral vertices of G is the monophonic periphery $P_m(G)$ of G.*

Example 2.13. Consider the graph G given in **Figure 1**. It is easily verified that v_5 is the monophonic central vertex and v_2, v_4, and v_6 are the monophonic peripheral vertices of G.

Remark 2.14. The monophonic center of a connected graph *need* not be connected. For the graph G given in **Figure 6**, $C_m(G) = \{v_3, v_6\}$.

Theorem 2.15. *Every graph is the monophonic center of some connected graph.*

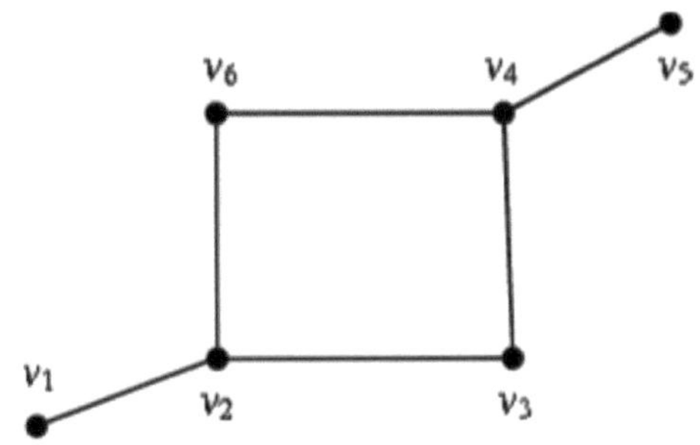

Figure 6. A graph G in Remark 2.14.

Proof. Let G be a graph. We show that G is the monophonic center of some graph. Let $l = d_m$ be the monophonic diameter of G. Let $P : u_1, u_2, \ldots, u_l$ and $Q : v_1, v_2, \ldots, v_l$ be two copies of the path P_l. The required graph H given in **Figure 7** is got from G, P, and Q by joining each vertex of G with u_1 in P and v_1 in Q. Then $e_{mH}(x) = d_m$ for each vertex x in G and $d_m + 1 \leq e_{mH}(x) \leq 2\, d_m$ for each vertex x not in G. Therefore, $V(G)$ is the set of monophonic central vertices of H and so $C_m(H) = G$.

More specifically, it is proved in Ref. [43] that the center of every connected graph G lies in a single block of G. Also, it is proved in Ref. [35] that the detour center of every connected graph G lies in a single block of G. The same result is true for the monophonic center also, as proved in the following theorem.

Theorem 2.16. *The monophonic center $C_m(G)$ of every connected graph G is a subgraph of some block of G.*

Proof. Suppose that there is a connected graph G such that its monophonic center $C_m(G)$ is not a subgraph of a single block of G. Then G has a cut vertex v such that $G - v$ contains two components H_1 and H_2, each containing vertices of $C_m(G)$. Let u be a vertex of G such that $e_m(v) = d_m(u,v)$, and let P_1 be a $u - v$ longest monophonic path in G. Then at least one of H_1 and H_2 contains no vertices of P_1, say H_2 contains no vertex of P_1. Now, take a vertex w in $C_m(G)$ that belongs to H_2, and let P_2 be a $v - w$ longest monophonic path in G. Since v is a cut vertex, P_1 followed by P_2 gives a $u - w$ longest monophonic path with its length greater than that of P_1. This gives $e_m(w) > e_m(v)$ so that w is not a monophonic central vertex of G, which is a contradiction.

Corollary 2.17. *For any tree, the monophonic center is isomorphic to K_1 or K_2.*

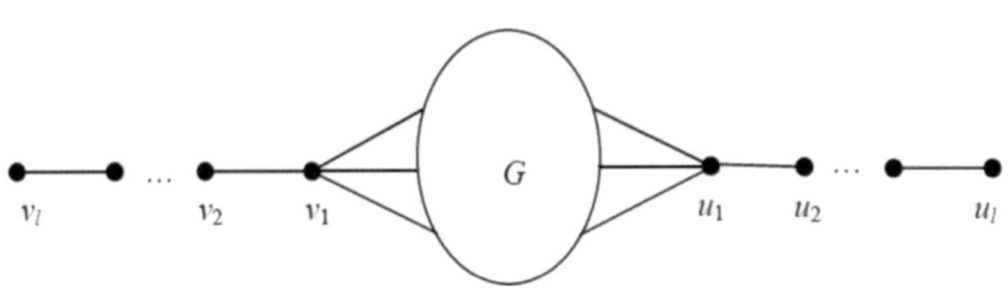

Figure 7. A graph H in Theorem 2.15.

It is proved in Ref. [44] that a nontrivial graph G is the periphery of some connected graph if and only if every vertex of G has eccentricity 1 or no vertex of G has eccentricity 1. Also, it is proved in Ref. [35] that a connected graph G of order $p \geq 3$ and radius 1 is the detour periphery of some connected graph if and only if G is Hamiltonian. A similar result is given in the next theorem, and for a proof, one may refer to Ref. [45].

Theorem 2.18. *A nontrivial graph G is the monophonic periphery of some connected graph if and only if every vertex of G has monophonic eccentricity* 1 *or no vertex of G has monophonic eccentricity* 1.

Definition 2.19. *A connected graph G is monophonic self-centered if* $rad_m\, G = diam_m\, G$*, that is, if G is its own monophonic center.*

Example 2.20. The complete graph K_n, the cycle C_n, and the complete bipartite graph $K_{m,n}$ $(m, n \geq 2)$ are monophonic self-centered graphs.

The following problem is left open.

Problem 2.21. *Characterize monophonic self-centered graphs.*

Further results on monophonic distance in graphs can be found in Refs. [45, 46].

3. Detour monophonic number

Throughout this section, by a $u - v$ detour monophonic path, we mean a longest $u - v$ monophonic path.

Definition 3.1. *A set S of vertices of a connected graph G is called a detour monophonic set if every vertex of G lies on a* $u - v$ *detour monophonic path for some* $u, v \in S$*. The detour monophonic number of G is defined as the minimum cardinality of a detour monophonic set of G and is denoted by* $dm(G)$.

Example 3.2. For the graph G given in **Figure 8**, $S_1 = \{v_1, v_2, v_3\}$, $S_2 = \{v_2, v_3, v_4\}$, $S_3 = \{v_5, v_6, v_2\}$, $S_4 = \{v_5, v_6, v_3\}$, $S_5 = \{v_1, v_3, v_5\}$, $S_6 = \{v_1, v_3, v_6\}$, $S_7 = \{v_2, v_4, v_5\}$, and $S_8 = \{v_2, v_4, v_6\}$ are the minimum detour monophonic sets of G and so $dm(G) = 3$.

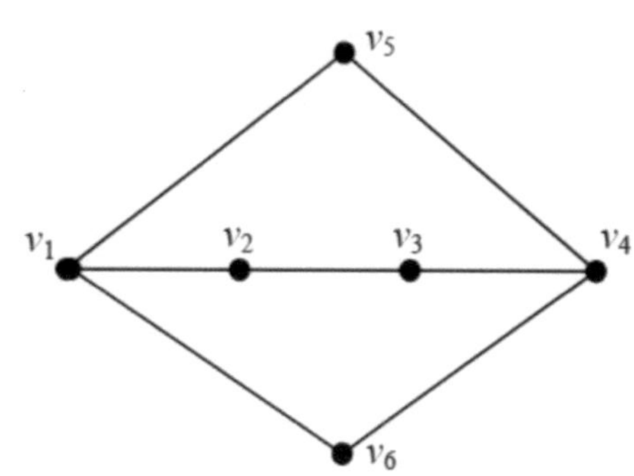

Figure 8. The graph G in Example 3.2.

If a vertex belongs to every minimum detour monophonic set of G, then it is called a *detour monophonic vertex* of G. If S is the unique minimum detour monophonic set of G, then S is the set of all detour monophonic vertices of G. In the next theorem, we show that there are certain vertices in a nontrivial connected graph G that are detour monophonic vertices of G.

Theorem 3.3. *Every detour monophonic set of a connected graph G contains all its extreme vertices. Moreover, if the set of all extreme vertices S of G is a detour monophonic set of G, then S is the unique minimum detour monophonic set of G.*

Proof. Let v be an extreme vertex and let S be a detour monophonic set of G. If v is not an element of S, then there exist two elements x and y in S such that v is an internal vertex of an $x-y$ detour monophonic path, say P. Let u and w be the vertices on P adjacent to v. Then u and w are not adjacent and so v is not an extreme vertex of G, which is a contradiction. Therefore v belongs to every detour monophonic set of G. Thus, if S is the set of all extreme vertices of G, then $dm(G) \geq |S|$. On the other hand, if S is a detour monophonic set of G, then $dm(G) \leq |S|$. Therefore $dm(G) = |S|$ and S is the unique minimum detour monophonic set of G.

The following two theorems are easy to prove.

Theorem 3.4. *Let G be a connected graph with a cut vertex v and let S be a detour monophonic set of G. Then every component of $G - v$ contains an element of S.*

Theorem 3.5. *No cut vertex of a connected graph G belongs to any minimum detour monophonic set of G.*

Since every end-block B is a branch of G at some cut vertex, it follows Theorem 3.4 and Theorem 3.5 that every minimum detour monophonic set of G contains at least one vertex from B that is not a cut vertex. Thus the following corollaries are consequences of Theorems 3.4 and 3.5.

Corollary 3.6. *If G is a connected graph with $k \geq 2$ end-blocks, then $dm(G) \geq k$.*

Corollary 3.7. *If k is the maximum number of blocks to which a vertex in a graph G belongs, then $dm(G) \geq k$.*

Theorem 3.8. *For any connected graph G, $2 \leq dm(G) \leq p$.*

Theorem 3.9. *For any connected graph G, $dm(G) = p$ if and only if G is complete.*

Proof. Let $dm(G) = p$. Suppose that G is not a complete graph. Then there exist two vertices u and v such that u and v are not adjacent in G. Since G is connected, there is a detour monophonic path from u to v, say P, with length at least 2. Clearly, $(V(G) - V(P)) \cup \{u, v\}$ is a detour monophonic set of G and hence $dm(G) \leq p - 1$, which is a contradiction. Conversely, if G is complete, then by Theorem 3.3, $dm(G) = p$.

Theorem 3.10. *If G is a nontrivial connected graph of order p and monophonic diameter d, then $dm(G) \leq p - d + 1$.*

Proof. Let $x, y \in V(G)$ such that G contains an $x - y$ detour monophonic path P of length $diam_m\, G = d$. Let $S = (V(G) - V(P)) \cup \{x, y\}$. Since S is a detour monophonic set of G, it follows that $dm(G) \leq |S| \leq p - d + 1$.

http://dx.doi.org/10.5772/intechopen.68668

Theorem 3.11. *For every nontrivial tree T of order p and monophonic diameter d, $dm(T) = p - d + 1$ if and only if T is a caterpillar.*

Proof. Let T be any nontrivial tree. Let $P : u = v_0, v_1, \ldots, v_d$ be a monophonic diametral path. Let k be the number of endvertices of T and let l be the number of internal vertices of T other than $v_1, v_2, \ldots, v_{d-1}$. Then $d - 1 + l + k = p$. By Theorem 3.3 and Theorem 3.5, $dm(T) = k$ and so $dm(T) = p - d - l + 1$. Hence $dm(T) = p - d + 1$ if and only if $l = 0$, if and only if all the internal vertices of T lie on the monophonic diametral path P, and if and only if T is a caterpillar.

It is known that $rad_m\ G \le diam_m\ G$ for a connected graph G. It is proved in Ref. [45] that if a and b are any two positive integers such that $a \le b$, then there is a connected graph G with $rad_m\ G = a$ and $diam_m\ G = b$. The same result can also be extended so that the detour monophonic number can be prescribed when $rad_m\ G < diam_m\ G$, and for a proof, one may refer to Ref. [47].

Theorem 3.12. *For positive integers r, d and $n \ge 4$ with $r < d$, there exists a connected graph G with $rad_m\ G = r$, $diam_m\ G = d$ and $dm(G) = n$.*

Problem 3.13. For any three positive integers r, d and $n \ge 4$ with $r = d$, does there exist a connected graph G with $rad_m\ G = r$, $diam_m\ G = d$ and $dm(G) = n$?

3.1. Upper detour monophonic number

Definition 3.14. *A detour monophonic set S of a connected graph G is called a minimal detour monophonic set if no proper subset of S is a detour monophonic set of G. The maximum cardinality of a minimal detour monophonic set of G is the upper detour monophonic number of G, denoted by $dm^+(G)$.*

Example 3.15. Consider the graph G given in **Figure 8**. The minimal detour monophonic sets are $S_1 = \{v_1, v_2, v_3\}$, $S_2 = \{v_2, v_3, v_4\}$, $S_3 = \{v_5, v_6, v_2\}$, $S_4 = \{v_5, v_6, v_3\}$, $S_5 = \{v_1, v_3, v_5\}$, $S_6 = \{v_1, v_3, v_6\}$, $S_7 = \{v_2, v_4, v_5\}$, $S_8 = \{v_2, v_4, v_6\}$ and $S_9 = \{v_1, v_4, v_5, v_6\}$. For this graph, the upper detour monophonic number is 4, and the detour monophonic number is 3.

Note 3.16. Every minimum detour monophonic set is a minimal detour monophonic set, but the converse is not true. For the graph G given in **Figure 8**, S_9 is a minimal detour monophonic set, but it is not a minimum detour monophonic set of G.

The following three theorems are easy to prove.

Theorem 3.17. *For any connected graph G, $2 \le dm(G) \le dm^+(G) \le p$.*

Theorem 3.18. *For a connected graph G, $dm(G) = p$ if and only if $dm^+(G) = p$.*

Theorem 3.19. *If G is a connected graph of order p with $dm(G) = p - 1$, then $dm^+(G) = p - 1$.*

The next theorem is an interesting realization result, and for a proof, one may refer to Ref. [48].

Theorem 3.20. *For any three positive integers a, b and n with $2 \le a \le n \le b$, there is a connected graph G with $dm(G) = a$, $dm^+(G) = b$ and a minimal detour monophonic set of cardinality n.*

3.2. Forcing detour monophonic number

A connected graph G may contain more than one minimum detour monophonic sets. For example, the graph G given in **Figure 8** contains eight minimum detour monophonic sets. For each minimum detour monophonic set S in G, there is always some subset T of S that uniquely determines S as the minimum detour monophonic set containing T. Such sets are called "forcing detour monophonic subsets" and these sets are discussed in this section.

Definition 3.21. *Let S be a minimum detour monophonic set of a connected graph G. A subset S' of S is a forcing detour monophonic subset for S if S is the unique minimum detour monophonic set that contains S'. A forcing detour monophonic subset for S of minimum cardinality is a minimum forcing detour monophonic subset of S. The cardinality of a minimum forcing detour monophonic subset of S is the forcing detour monophonic number $fdm(S)$ in G. The forcing detour monophonic number of G is $fdm(G) = \min\{fdm(S)\}$, where the minimum is taken over all minimum detour monophonic sets S in G.*

Example 3.22. *For the graph G given in **Figure 9**, $S_1 = \{z, w, v\}$, $S_2 = \{z, w, u\}$ and $S_3 = \{z, w, x\}$ are the minimum detour monophonic sets of G. It is clear that $fdm(S_1) = 1$, $fdm(S_2) = 1$ and $fdm(S_3) = 1$ so that $fdm(G) = 1$. For the graph G given in **Figure 10**, $S = \{y, v\}$ is the unique minimum detour monophonic set of G and so $fdm(G) = 0$.*

The next theorem follows immediately from the definitions of the detour monophonic number and the forcing detour monophonic number of a graph G.

Theorem 3.23. *For a connected graph G, $0 \le fdm(G) \le dm(G) \le p$.*

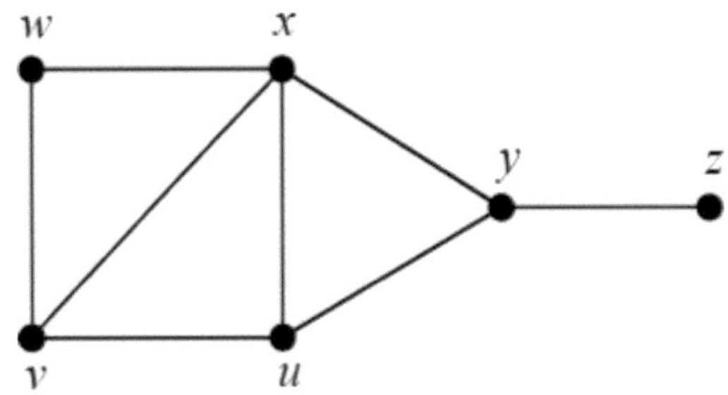

Figure 9. A graph G with $fdm(G) = 1$.

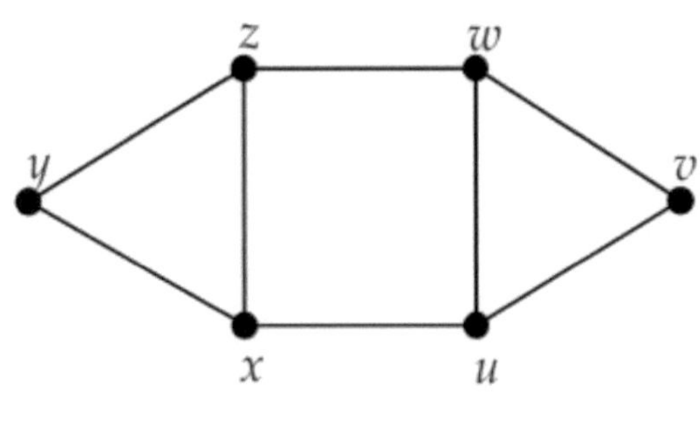

Figure 10. A graph G with $fdm(G) = 0$.

The following theorem characterizes graphs G for which $fdm(G) = 0$, $fdm(G) = 1$ and $fdm(G) = dm(G)$. The proof is an easy consequence of the definitions of the detour monophonic number and the forcing detour monophonic number.

Theorem 3.24. *Let G be a connected graph. Then*

(i) *$fdm(G) = 0$ if and only if G contains exactly one minimum detour monophonic set.*

(ii) *$fdm(G) = 1$ if and only if G contains two or more minimum detour monophonic sets, one of which is a unique minimum detour monophonic set that contains one of its elements.*

(iii) *$fdm(G) = dm(G)$ if and only if no minimum detour monophonic set of G is the unique minimum detour monophonic set that contains any of its proper subsets.*

The next theorem gives a realization result for the parameters $fdm(G)$ and $dm(G)$, and for a proof, the reader may refer to Ref. [49].

Theorem 3.25. *For every pair a, b of positive integers with $0 \leq a < b$ and $b \geq 2$, there exists a connected graph G such that $fdm(G) = a$ and $dm(G) = b$.*

Further results on detour monophonic concepts in graphs can be found in Refs. [47–50].

4. Vertex detour monophonic number

The parameter detour monophonic number of a graph is global in the sense that there is exactly one detour monophonic number for a graph. The concept of detour monophonic sets and detour monophonic numbers by fixing a vertex of a graph was also introduced and discussed in this section. With respect to each vertex of a graph, there is a detour monophonic number, and so there will be at most as many detour monophonic numbers as there are vertices in the graph.

Definition 4.1. *For any vertex x in a connected graph G, a set S_x of vertices in G is called an x-detour monophonic set if every vertex of G lies on an $x - y$ detour monophonic path in G for some y in S_x. The x-detour monophonic number of G, denoted by $dm_x(G)$, is defined to be the minimum cardinality of an x-detour monophonic set of G. An x-detour monophonic set of cardinality $dm_x(G)$ is called a dm_x-set of G.*

It is easy to observe that for any vertex x in G, x does not belong to any dm_x-set of G.

Example 4.2. For the graph G given in **Figure 11**, the minimum vertex detour monophonic sets and the vertex detour monophonic numbers are given in **Table 3**.

The next two theorems are easy to prove.

Theorem 4.3. *For any vertex x in a connected graph G, the following results hold.*

(i) *Every dm_x-set of G contains all its extreme vertices other than the vertex x (whether x is extreme vertex or not).*

(ii) *No dm_x-set of G contains a cut vertex of G.*

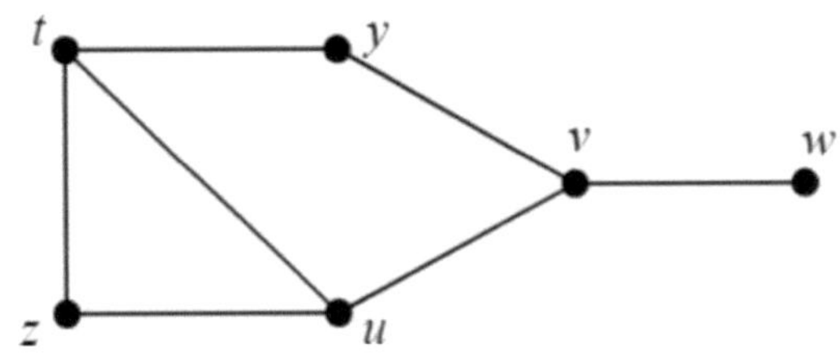

Figure 11. The graph G in Example 4.2.

Vertex	Minimum vertex detour monophonic sets	Vertex detour monophonic number
t	$\{z,w\}$	2
y	$\{w,z,t\}$, $\{w,z,u\}$	3
z	$\{u,w\}$, $\{w,y\}$	2
u	$\{w,z,y\}$	3
v	$\{w,t,z\}$, $\{w,u,z\}$	3
w	$\{t,z\}$, $\{z,u\}$	2

Table 3. Vertex detour monophonic numbers of the graph G in Figure 11.

Theorem 4.4. *For any vertex x in a connected graph G of order p,* $1 \le dm_x(G) \le p - 1$.

Theorem 4.5. *For any vertex x in a connected graph G of order p,* $dm_x(G) = p - 1$ *if and only if* $deg\ x = p - 1$.

Proof. Let x be any vertex in a connected graph G of order p. Let $dm_x(G) = p - 1$. If $deg\ x < p-1$, then there is a vertex u in G that is not adjacent to x. Since G is connected, there is a detour monophonic path from x to u, say P, with length greater than or equal to 2. Then $(V(G)-V(P)) \cup \{u\}$ is an x-detour monophonic set of G so that $dm_x(G) \le p - 2$, which is a contradiction. Conversely, let $deg\ x = p - 1$. Hence x is adjacent to all other vertices of G. This shows that all these vertices form the dm_x-set of G and so $dm_x(G) = p - 1$.

Corollary 4.6. *A graph G is complete if and only if* $dm_x(G) = p - 1$ *for every vertex x in G.*

4.1. Upper vertex detour monophonic number

Definition 4.7. *Let x be any vertex of a connected graph G. An x-detour monophonic set* S_x *is called a minimal x-detour monophonic set if no proper subset of* S_x *is an x-detour monophonic set. The upper x-detour monophonic number is the maximum cardinality of a minimal x-detour monophonic set of G and is denoted by* $dm_x^+(G)$.

Example 4.8. For the graph G given in **Figure 12**, the minimum vertex detour monophonic sets the vertex detour monophonic numbers, the minimal vertex detour monophonic sets and the upper vertex detour monophonic numbers are given in **Table 4**.

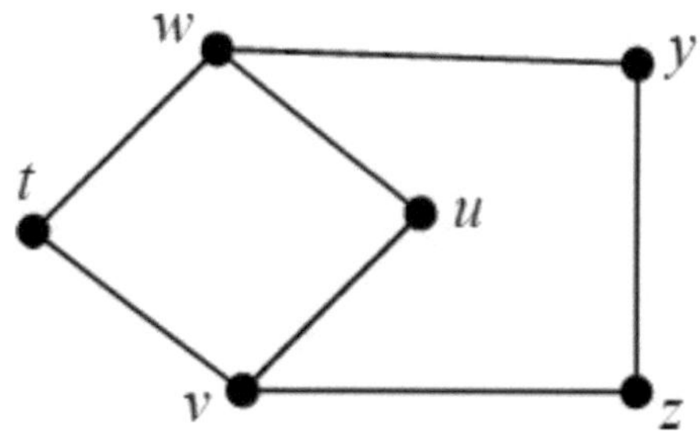

Figure 12. The graph G in Example 4.8.

Vertex x	Minimum x-detour monophonic sets	$dm_x(G)$	Minimal x-detour monophonic sets	$dm_x^+(G)$
t	$\{u,y\}$, $\{u,z\}$	2	$\{u,y\}$, $\{u,z\}$	2
u	$\{t,y\}$, $\{t,z\}$	2	$\{t,y\}$, $\{t,z\}$	2
v	$\{w,y\}$, $\{z,y\}$	2	$\{w,y\}$, $\{z,y\}$, $\{w,t,u\}$	3
w	$\{z,y\}$, $\{z,v\}$	2	$\{z,y\}$, $\{z,v\}$, $\{v,t,u\}$	3
y	$\{v,z\}$, $\{v,t\}$, $\{v,u\}$	2	$\{v,z\}$, $\{v,t\}$, $\{v,u\}$, $\{t,u,w\}$	3
z	$\{w,y\}$, $\{w,t\}$, $\{w,u\}$	2	$\{w,y\}$, $\{w,t\}$, $\{w,u\}$, $\{v,t,u\}$	3

Table 4. Upper vertex detour monophonic numbers of the graph G in Figure 12.

Since every minimum x-detour monophonic set is a minimal x-detour monophonic set, we have $1 \leq dm_x(G) \leq dm_x^+(G) \leq p - 1$. In view of this, we have the following theorems, and for proofs one may refer to Ref. [51].

Theorem 4.9. *Let x be any vertex in a connected graph G of order $p \geq 3$. If $dm_x(G) = 1$, then $dm_x^+(G) \leq p - 2$.*

Theorem 4.10. *Let x be any vertex in a connected graph G. Then $dm_x(G) = p - 1$ if and only if $dm_x^+(G) = p - 1$.*

Theorem 4.11. *For any four integers j, k, l and p with $2 \leq j \leq k \leq l \leq p - 7$, there exists a connected graph G of order p with $dm_x(G) = j$, $dm_x^+(G) = l$ and a minimal x-detour monophonic set of cardinality k.*

4.2. Forcing vertex detour monophonic number

Definition 4.12. *Let x be any vertex of a connected graph G and let S_x be a minimum x-detour monophonic set of G. A subset S' of S_x is an x-forcing subset for S_x if S_x is the unique minimum x-detour monophonic set that contains S'. An x-forcing subset for S_x of minimum cardinality is a minimum x-forcing subset of S_x. The cardinality of a minimum x-forcing subset of S_x is the forcing x-detour monophonic number $fdm_x(S_x)$ in G. The forcing x-detour monophonic number of G is $fdm_x(G) = \min\{fdm_x(S_x)\}$, where the minimum is taken over all minimum x-detour monophonic sets S_x in G.*

Definition 4.13. *Let x be any vertex of a connected graph G. The upper forcing x-detour monophonic number,* $fdm_x^+(G)$*, of G is the maximum forcing x-detour monophonic number among all minimum x-detour monophonic sets of G.*

Example 4.14. For the graph G given in **Figure 13**, the minimum vertex detour monophonic sets, the vertex detour monophonic numbers, the forcing vertex detour monophonic sets, the forcing vertex detour monophonic numbers and the upper forcing vertex detour monophonic numbers are given in **Table 5**.

Theorem 4.15. *For any vertex x in a connected graph G,* $0 \leq fdm_x(G) \leq fdm_x^+(G) \leq dm_x(G)$.

The following theorem gives a realization result for the parameters $fdm_x(G)$, $fdm_x^+(G)$, $dm_x(G)$, and for a proof, one may refer to Ref. [52].

Theorem 4.16. *For any three integers r, s and t with* $2 \leq r \leq s \leq t$ with $2r - s \geq 0$, *there exists a connected graph G with* $fdm_x(G) = r$, $fdm_x^+(G) = s$ *and* $dm_x(G) = t$ *for some vertex x in G.*

There are useful applications of these concepts to security-based communication network design. In the case of designing the channel for a communication network, although all the vertices are covered by the network when considering detour monophonic sets, some of the edges may be left out. This drawback is rectified in the case of edge detour monophonic sets so that considering edge detour monophonic sets is more advantageous to real-life application of communication networks. The edge detour monophonic sets are discussed in Refs. [53–55].

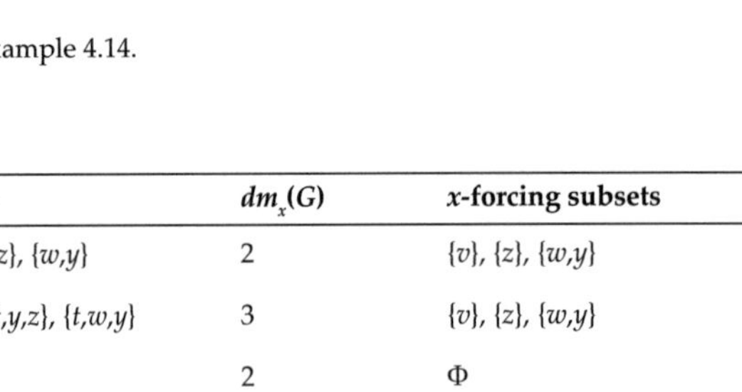

Figure 13. The graph G in Example 4.14.

Vertex x	dm_x-sets	$dm_x(G)$	x-forcing subsets	$fdm_x(G)$	$fdm_x^+(G)$
t	$\{v,w\}$, $\{y,z\}$, $\{w,y\}$	2	$\{v\}$, $\{z\}$, $\{w,y\}$	1	2
u	$\{t,v,w\}$, $\{t,y,z\}$, $\{t,w,y\}$	3	$\{v\}$, $\{z\}$, $\{w,y\}$	1	2
v	$\{t,z\}$	2	Φ	0	0
w	$\{t,v\}$, $\{t,z\}$	2	$\{v\}$, $\{z\}$	1	1
y	$\{t,v\}$, $\{t,z\}$	2	$\{v\}$, $\{z\}$	1	1
z	$\{t,v\}$	2	Φ	0	0

Table 5. Forcing and upper forcing vertex detour monophonic numbers of the graph G in Figure 13.

5. Conclusion

In this chapter, the new distance known as monophonic distance in a graph is introduced, and its properties are studied. Its relationship with the usual distance and detour distance is discussed. Various realization theorems are proved with regard to the radius (diameter), monophonic radius (monophonic diameter) and detour radius (detour diameter). Results regarding monophonic center and monophonic periphery of a graph are presented. Further, the concept of a detour monophonic set in a graph is introduced and its various properties are presented. Consequently, the parameters, viz., detour monophonic number, upper detour monophonic number and forcing detour monophonic number of a graph are introduced and studied. In a similar way, the vertex detour monophonic number, the upper vertex detour monophonic number and the forcing vertex detour monophonic number of a graph are introduced and studied. Many interesting characterization theorems and also realization theorems with regard to all these parameters are presented. The results presented in this chapter would help the researchers in graph theory to develop new results and applications to various branches of science.

Author details

P. Titus[1]* and A.P. Santhakumaran[2]

*Address all correspondence to: titusvino@yahoo.com

1 Department of Mathematics, University College of Engineering Nagercoil, Anna University, Tirunelveli Region, Nagercoil, India

2 Department of Mathematics, Hindustan Institute of Technology and Science, Chennai, India

References

[1] Bandelt HJ, Mulder HM. Distance-hereditary graphs. Journal of Combinatorial Theory, Series B. 1986;**41**:182-208

[2] Buckley F, Harary F. Distance in Graphs. Redwood City, CA: Addison-Wesley; 1990

[3] Buckley F, Lewinter M. A note on graphs with diameter-preserving spanning trees. Journal of Graph Theory. 1988;**12**(4):525-528

[4] Floyd RW. Algorithm 97: Shortest path. Communications of the ACM. 1962;**5**:345

[5] Howorka E. A characterization of distance-hereditary graphs. The Quarterly Journal of Mathematics, Oxford. 1977;**28**:417-420

[6] Ostrand PA. Graphs with specified radius and diameter. Discrete Mathematics. 1973;**4**: 71-75

[7] Plesnık J. On the sum of all distance in a graph or digraph. Journal of Graph Theory. 1984;**8**:1-24

[8] Swart CS. Distance measures in graphs and subgraphs [thesis]. Durban: University of Natal; 1996

[9] Vizing VG. On the number of edges in a graph with a given radius. Doklady Akademii Nauk SSSR. 1967;**173**:1245-1246

[10] Buckley F, Miller Z, Slater PJ. On graphs containing a given graph as center. Journal of Graph Theory. 1981;**5**:427-432

[11] Freeman LC. Centrality in social networks: Conceptual clarification. Social Networks. 1978/1979;**1**:215-239

[12] Harary F, Melter RA. On the metric dimension of a graph. Ars Combinatoria. 1976;**2**:191-195

[13] Kouider M, Winkler P. Mean distance and minimum degree. Journal of Graph Theory. 1997;**25**:95-99

[14] Mulder HM. n-cubes and median graphs. Journal of Graph Theory. 1980;**4**:107-110

[15] Santhakumaran AP. Median with respect to cliques. Acta Ciencia Indica Mathematics. 2005;**31**(3):633-642

[16] Santhakumaran AP. Periphery with respect to cliques in graphs. Journal of Discrete Mathematical Sciences & Cryptography. 2007;**10**(2):245-254

[17] Santhakumaran AP. Center of a graph with respect to edges. SCIENTIA Series A: Mathematical Sciences. 2010;**19**:13-23

[18] Santhakumaran AP. Median of a graph with respect to edges. Discussiones Mathematicae Graph Theory. 2012;**32**:19-29

[19] Slater PJ. Centers to centroids in graphs. Journal of Graph Theory. **2**(3):209-222

[20] Slater PJ. Centrality of paths and vertices in a graph: Cores and pits. The Theory and Applications of Graphs. Wiley; 1981. pp. 529-542

[21] Winkler P. Mean distance and the four thirds conjecture. Congressus Numerantium. 1986;**54**:53-61

[22] Zelinka B. Medians and peripherians of trees. Archiv der Mathematik. 1968;**4**:87-95

[23] Chartrand G, Harary F, Swart H, Zhang P. Geodomination in graphs. Bulletin of the Institute of Combinatorics and its Applications. 2001;**31**:51-59

[24] Chartrand G, Harary F, Zhang P. On the geodetic number of a graph. Networks. 2002;**39**:1-6

[25] Chartrand G, Johns GL, Zhang P. On the detour number and geodetic number of a graph. Ars Combinatoria. 2004;**72**:3-15

[26] Santhakumaran AP, Jebaraj T. The upper double geodetic number of a graph. Malaysian Journal of Sciences. 2011;**30**(3):225-229

[27] Santhakumaran AP, Jebaraj T. Double geodetic number of a graph. Discussiones Mathematicae Graph Theory. 2012;**32**:109-119

[28] Santhakumaran AP, Kumari Latha T. On the open geodetic number of a graph. SCIENTIA Series A: Mathematical Sciences. 2010;**19**:131-142

[29] Santhakumaran AP, Titus P. The upper connected geodetic number and forcing connected geodetic number of a graph. Discrete Applied Mathematics. 2009;**157**:1571-1580

[30] Santhakumaran AP, Ullas Chandran SV. The geodetic number of strong product graphs. Discussiones Mathematicae Graph Theory. 2010;**30**(4):687-700

[31] Santhakumaran AP, Ullas Chandran SV. The 2-edge geodetic number and graph operations. Arabian Journal of Mathematics. 2012;**I**:241-249

[32] Santhakumaran AP, Ullas Chandran SV. The k-edge geodetic number of a graph. Utilitas Mathematica. 2012;**88**:119-137

[33] Buckley F, Harary F, Quintas LV. Extremal results on the geodetic number of a graph. Scientia A. 1988;**2**:17-26

[34] Santhakumaran AP, Titus P. Geodesic graphs. Ars Combinatoria. 2011;**99**:75-82

[35] Chartrand G, Escuadro H, Zhang P. Detour distance in graphs. Journal of Combinatorial Mathematics and Combinatorial Computing. 2005;**53**:75-94

[36] Chartrand G, Zhang P. Distance in graphs – Taking the long view. AKCE International Journal of Graphs and Combinatorics. 2004;**1**(1):1-13

[37] Chartrand G, Johns GL, Zhang P. The detour number of a graph. Utilitas Mathematica. 2003;**64**:97-113

[38] Santhakumaran AP, Titus P. The vertex detour number of a graph. AKCE International Journal of Graphs and Combinatorics. 2007;**4**(1):99-112

[39] Santhakumaran AP, Titus P. The upper and forcing vertex detour numbers of a graph. International Journal of Mathematical Combinatorics. 2008;**2**:109-120

[40] Santhakumaran AP, Titus P. On the vertex geodomination number of a graph. Ars Combinatoria. 2011;**101**:137-151

[41] Santhakumaran AP, Titus P. The upper connected vertex detour number of a graph. Filomat. 2012;**26**(2):379-388. DOI: 10.2298/FIL1202379S

[42] Santhakumaran AP, Titus P. The geo-number of a graph. Ars Combinatoria. 2012;**106**: 65-78

[43] Harary F, Norman RZ. The dissimilarity characteristics of Husimi trees. Annals of Mathematics. 1953;**58**:134-141

[44] Bielak H, Syslo MM. Peripheral vertices in graphs. Studia Scientiarum Mathematicarum Hungarica. 1983;**18**:269-275

[45] Santhakumaran AP, Titus P. Monophonic distance in graphs. Discrete Mathematics, Algorithms and Applications. 2011;**3**(2):159-169

[46] Santhakumaran AP, Titus P. A note on "Monophonic distance in graphs". Discrete Mathematics, Algorithms and Applications. 2012;**4**(2):12500189

[47] Titus P, Ganesamoorthy K. On the detour monophonic number of a graph. Ars Combinatoria. 2016;**129**:33-42

[48] Titus P, Ganesamoorthy K. Upper detour monophonic number of a graph. Electronic Notes in Discrete Mathematics. 2016;**53**:331-342

[49] Titus P, Ganesamoorthy K. Forcing detour monophonic number of a graph. Facta Universitatis, Series: Mathematics and Informatics. 2013;**28**(2):211-220

[50] Titus P, Ganesamoorthy K, Balakrishnan P. The detour monophonic number of a graph. Journal of Combinatorial Mathematics and Combinatorial Computing. 2013;**84**:179-188

[51] Titus P, Balakrishnan P. The upper vertex detour monophonic number of a graph. Ars Combinatoria, to appear

[52] Titus P, Balakrishnan P. The forcing vertex detour monophonic number of a graph. AKCE International Journal of Graphs and Combinatorics. 2016; DOI: 10.1016/j.akcej.2016.03.002

[53] Santhakumaran AP, Titus P, Ganesamoorthy K, Balakrishnan P. Edge detour monophonic number of a graph. Proyecciones Journal of Mathematics. 2013;**32**(2):181-196

[54] Titus P, Balakrishnan P, Santhakumaran AP, Ganesamoorthy K. Connected edge detour monophonic number of a graph. Proceedings of the Jangjeon Mathematical Society. 2013;**16**(4):517-528

[55] Titus P, Ganesamoorthy K. Upper edge detour monophonic number of a graph. Proyecciones Journal of Mathematics. 2014;**33**(2):175-187

Chapter 7

Spectra and Quantum Transport on Graphs

Victor Chulaevsky

Additional information is available at the end of the chapter

http://dx.doi.org/10.5772/intechopen.68480

Abstract

This chapter is devoted to various interactions between the graph theory and mathematical physics of disordered media, studying spectral properties of random quantum Hamiltonians. We show how the notions, methods, and constructions of graph theory can help one to solve difficult problems, and also highlight recent developments in spectral theory of multiparticle random Hamiltonians which both benefit from graph-theoretical methods and suggest original structures where new insights are required from various areas of mathematical physics in a broad sense.

Keywords: isoperimetric estimates, Cheeger bound, Lifshitz tails, Anderson localization, multiparticle localization, quantum graphs

1. Introduction

The proposed chapter focuses on the methods and applications of the graph theory in the area of quantum transport on combinatorial and metric (often referenced to as "quantum") graphs. It is well known that perfectly periodic potentials in Euclidean spaces or on periodic lattices create favorable conditions for nonlocalized solutions of Schrödinger and/or wave equations. In quantum physics, this results in transport of quantum particles, for example, electrons or phonons. However, after the seminal, Nobel prize winning work [1] published by Philip W. Anderson in 1958, it has been realized by physicists that the propagation of quantum particles in an imperfect environment, modeled by a random or almost periodic electrostatic or magnetic potential, can be significantly inhibited, to the point where mobile quantum particles, e.g., electrons, are localized: their wave functions decay exponentially away from some loci— their respective localization centers. In many applications, the media where quantum particles propagate are not periodic crystals, but have instead a structure of more or less complex graphs formed by atoms, and therefore are treated as disordered media. The structural disorder can be complemented by a parametric one, e.g., in the context of weighted graphs where the canonical graph Laplacian Δ_Γ on Γ is modulated by variable weights assigned to the edges. In general terms of the spectral theory of unordered structures, this is an instance of the "off-diagonal" disorder. Furthermore, it is important to analyze the spectral properties of the discrete analogs of the Schrödinger operators $-\Delta_\Gamma + V$ on a graph Γ, where V is a fixed or random real-valued function on Γ.

The recent wake of interest to the nanosystems and molecular devices attracted many researchers to such models, and numerous intriguing problems in this area still remain challenging and wide open. It has been understood that the classical aspects of the graph theory, such as isoperimetric estimates (particularly, the Cheeger bounds) and deep results of the spectral theory of graphs, are of great importance to the localization/delocalization processes on graphs other than periodic lattices embedded in a Euclidean space.

Furthermore, the most recent developments in the spectral theory of disordered quantum systems, initiated independently and simultaneously by physicists (cf., e.g., [2]) and mathematicians (cf. [3–6], emphasized the role of interparticle interaction which had been consciously and explicitly neglected in the pioneering works due to the complexity of the analysis involved, although P. W. Anderson himself was concerned about possible effects of interaction on the fundamental properties of the quantum transport. Following the first mathematical works in this direction, the notion of a multiparticle quantum graph has been recently introduced by Sabri [7].

Summarizing, the proposed chapter provides to the reader an overview of synthetic techniques and results where the traditional problem of the combinatorial and spectral graph theory is intertwined with complementary structures, ideas, and methods of functional analysis and quantum mechanics, in response to the new challenges in modern technology.

2. Isoperimetric bounds, spectral gaps, and quantum localization

The integer lattices $\mathbb{Z}^d$, $d \geq 1$, endowed with the usual graph structure constitute a very particular class of connected graphs. The spectra and (generalized) eigenfunctions of their canonical graph Laplacians are easy to find, due to the commutative group structure of these lattices. The lattice Laplacian $\Delta_{\mathbb{Z}^d}$ is the canonical Laplacian on $\mathbb{Z}^d$ endowed with the graph structure where the edges are formed by the pairs (x, y) with Euclidean distance $|x - y|_2 = 1$. It follows from the Fourier analysis on the additive group $\mathbb{Z}^d$ that the spectral measure of $\Delta_{\mathbb{Z}^d}$ is absolutely continuous (with respect to the Lebesgue measure). The graph Laplacians Δ_Γ are defined as nonpositive operators, but in mathematical physics one considers $-\Delta_\Gamma$ instead. The spectrum of $-\Delta_{\mathbb{Z}^d}$ is easily computed, knowing that its Fourier image is the operator of multiplication by the function

$$\mathbf{p} = (p_1, \dots, p_d) \mapsto \sum_{j=1}^{d} (1 - \cos p_j). \tag{1}$$

The generalized eigenfunctions are given by the respective Fourier harmonics (plane waves) $\mathbf{x} \mapsto \exp\big(i(\mathbf{p}, \mathbf{x})\big)$, where $(\mathbf{p}, \mathbf{x}) = p_1 x_1 + \dots + p_d x_d$. In physics, one often works with finite cubes, where the eigenfunctions of the respective Laplacian are combinations of plane waves. Particular questions concerning the Laplacians relative to such finite graphs depend upon specific intended applications. A number of situations give rise to the quantitative analysis of a few of the lowest eigenvalues of $(-\Delta_\Gamma)$, numbered in increasing order. In particular, the spectral gap and the analytic form of the eigenfunctions with eigenvalues λ_0 and λ_1, known explicitly for the rectangles, are more difficult to analyze for general graphs. One possible question is about the size of the gap $\lambda_1 - \lambda_0$: it features a power-law decay as the size L of the cube $[1, L]^d$ grows, but what can one say about the spectral gap for less regular graphs? Furthermore, it is readily seen that the eigenvalue λ_1 is degenerate on the cube, and has multiplicity equal to the dimension d: the respective eigenfunctions are the lowest-frequency harmonics, related to the global geometry of the cube, but the situation for general graphs is more complex.

Before going to the answers, provided by the graph theory for a large class of nonperiodic graphs, we give some motivations coming from the spectral theory of random operators.

One remarkable phenomenon relative to disordered media was discovered in the 1960s by physicist I.M. Lifshitz [8] and colorfully called in the physics and mathematics communities "Lifshitz tails": the eigenvalue distribution of the random operators $H_V = -\Delta + V(\omega)$ decays extremely as the energy E approaches the lower edge of spectrum.

In this chapter, we always assume the random potential to take i.i.d. (independent and identically distributed) values. In addition, we assume the common probability distribution of these random values to admit a bounded probability density.

We shall use the following notions and notations. Given a potential $V : \mathbb{Z}^d \to \mathbb{R}$, we consider the discrete Schrödinger operator $H_V = -\Delta + V$, where Δ is the graph Laplacian on the integer lattice $\mathbb{Z}^d$. Further, for each integer $L \geq 1$ and a lattice point x denoted by $B(L, x)$ the cube centered at x of side length $2L$, and let H_L be the Schrödinger operator $-\Delta_{B(L,0)} + V$ in the cube $B(L, 0)$; here, $-\Delta_{B(L,0)}$ is the canonical graph Laplacian in $B(L, 0)$ with the graph structure inherited from the integer lattice, where the edges are given by the pairs of nearest neighbours in the Euclidean distance. Next, denote by λ_k the eigenvalues of H_L numbered in increasing order, and introduce the finite-volume eigenvalue counting function

$$N(E, L) = \frac{1}{|B(L, 0)}\mathrm{card}\{k : \lambda_k \leq E\}, E \in \mathbb{R}. \tag{2}$$

Definition 1. *The limiting eigenvalue distribution function* $N(E)$ *of the operator* H_V*is the limit*

$$N(E) := lim_{L\to\infty} N(E, L), \tag{3}$$

whenever it exists. Otherwise, we say that the limiting distribution function does not exist.

In the case where a fixed potential $V : \mathbb{Z}^d \to \mathbb{R}$ is replaced by a sample $V(\omega)$ of a random field on the lattice, the operator $H_V = H_V(\omega)$ also becomes random.

In physical terminology, widely used also in mathematical physics of disordered media, $N(E)$ is usually called the integrated density of states (IDS). The existence of the above limit is not obvious and, generally speaking, the limit may not exist. However, the existence of $N(E)$ for any energy can be established by the methods of ergodic theory in a particular, but very rich and useful for physical applications class of ergodic operators (including all i.i.d. potentials), as well as for the periodic and almost-periodic potentials. In fact, the latter classes can be incorporated into the general scheme of ergodic operators; cf., e.g., the monographs [9, 10]. Moreover, the IDS for ergodic potentials is nonrandom; in physical terminology, IDS is a "self-averaging" quantity. Simply put, spatial average coincides with the ensemble average for ergodic operators.

Whenever the potential V of the Schrödinger operator $H_V = -\Delta + V$ is lower bounded, e.g., nonnegative, H_V, and its spectrum $\mathrm{Spec}(H_V)$ have the same property, since $-\Delta$ is nonnegative. Therefore, $E_0 := \inf\ \mathrm{Spec}(H_V) > -\infty$. In physics, E_0 is called the ground state energy. A number of important quantities and phenomena are related to the ground state energy and also to the behavior of the IDS as the energy E approaches E_0. Lifshitz [8] discovered that for a large class of Hamiltonians with random potential energy, including random Schrödinger operators $H_V = -\Delta + V(\omega)$ on a lattice, the IDS decays very fast as $E \downarrow E_0$: for some $C_1, C > 0$

$$N(E) \sim C_1 \exp\left(-C(E - E_0)^{\frac{-d}{2}}\right). \tag{4}$$

Lifshitz tails have numerous important ramifications in theoretical and experimental physics. They also result in a nonperturbative onset of the Anderson localization on lattices for any, arbitrarily small amplitude of the random potential $V(\omega)$. Away from the spectral edge, the proofs of localization require a sufficiently large amplitude of $V(\omega)$; moreover, it is widely believed that in dimension $d \geq 3$, in the models where the random potential $gV(\omega)$ has a sufficiently small amplitude $|g|$, there are intervals I of energy where the corresponding generalized eigenfunctions ("extended quantum states") are not square-summable, and the spectral measure has in I a nontrivial absolutely continuous component. In simpler terms, there is a nontrivial quantum transport in some energy zones.

A substantial progress has been achieved in the direction of proofs of localization near the spectral edge (or edges). For a long time, most of the efforts have been made in the analysis of lattice Hamiltonians $H_V = -\Delta + V(\omega)$. Recently, it has been shown in Ref. [11] that a number of results obtained on the integer lattices can be extended to much more general graphs of polynomial (or, more generally, subexponential) growth. The key point is the availability of lower bounds on the spectral gap in terms of the Cheeger's constant of the graph.

Consider a lattice cube $B = B(L, 0)$ with the graph structure inherited from the lattice, and the random lattice Schrödinger operator $H_{B,V}(\omega) = -\Delta_B + V(\omega)$ on it. The starting point of the localization analysis of this finite-volume Hamiltonian is the estimate of the probability to have some eigenvalues of $H_{B,V}(\omega)$ in a small interval $I_\epsilon = [E_0, E_0 + \epsilon]$ near the spectral edge E_0. This is closely related to the Lifshitz tails. One needs a finite-volume analog of the limiting Lifshitz asymptotics, but for the localization analysis, one can settle for a sufficiently strong, albeit not necessarily sharp, upper bound on the probability

$$\mathbf{P}\Big(\exists\, E_j \in \mathrm{Spec}\Big(H_{B,V}(\omega)\Big) \,:\, E_j \in I_\epsilon\Big). \tag{5}$$

Now recall Temple's inequality [12].

Proposition 1. *Let A be a self-adjoint operator in a finite-dimensional Hilbert space $\mathcal{H}$, and $E_0 = \inf \mathrm{Spec}(A)$ be a simple eigenvalue. If a vector ψ with unit norm satisfies $\langle\psi, A\psi\rangle < E_1 := \inf \mathrm{Spec}(A) \setminus \{E_0\}$, then*

$$E_0 \geq \langle\psi, A\psi\rangle - \frac{\langle\psi, A^2\psi\rangle - (\langle\psi, A\psi\rangle)^2}{E_1 - \langle\psi, A\psi\rangle}.$$

Now one can see the importance of the size of the lowest spectral gap, $\eta = E_1 - E_0$. As was mentioned above, η is easily calculated explicitly for the Laplacians in lattice rectangles, but of course there is no universal formula for general finite graphs. The following result was obtained in Ref. [11].

Proposition 2. *Let be given a finite connected subgraph G of a locally finite countable connected graph Γ satisfying the following condition: there exists some real constants $d \geq 1$ and $C > 0$ such that any ball $B(L, x) \subset \Gamma$ of radius $L \geq 1$ has cardinality*

$$|B(L, x)| \leq CL^d. \tag{6}$$

Then

$$E_1(-\Delta_G) \geq c_d |G|^{-2}, \quad c_d > 0.$$

Apart from the canonical negative Laplacian $(-\Delta_G)$, it is often more convenient to work with its modified variant $\mathcal{L}_G$ defined by

$$(\mathcal{L}_G\ f)(x) = (n_G(x))^{-1/2}\ \ ((n_G(x))^{-1/2}\ \ f(x) - (n_G(y))^{-1/2} f(y)) \tag{7}$$

where $(n_G(x)$ and $(n_G(y)$ are the coordination numbers of the vertices x and y, respectively: $n_G(x) = \mathrm{card}\Big(B(1, x) \setminus \{x\}\Big)$. The coordination numbers are nonzero, whenever the graph is connected and has more than one vertex. Below we quote the bounds obtained for the modified Laplacian, but, up to some constants, they remain valid for the original Laplacian.

Definition 2. *The Cheeger's constant of a finite connected graph G is the following quantity:*

$$h(G) := \min_{G = W \sqcup W^c} \frac{|\partial W|}{\min[\mathrm{vol}(W), \mathrm{vol}(W^c)]},$$

where the minimum is taken over all nontrivial partitions $G = W \sqcup W^c$ of the graph G into disjoint subgraphs W and its complement W^c.

Denote by $\mu_k(G)$, $k \geq 0$, the eigenvalues of $\mathcal{L}_G$ numbered, as their counterparts $\lambda_k(G)$, in increasing order. Then one has the following results (cf. [13]).

Proposition 3. Let be given a finite connected graph G of diameter $D_G := \mathrm{diam} G$. Let $\mu_1(G)$ be the first nonz ero eigenvalue of the modified graph Laplacian $\mathcal{L}_G$, and $h(G)$ the Cheeger's constant of G. Then

$$\min\left[\frac{\left(h(G)\right)^2}{2}, \frac{1}{D_G \mathrm{vol}(G)}, \frac{2}{\left(\mathrm{vol}(G)\right)^2}\right] \le \mu_1(G) \le 2h(G). \tag{8}$$

For a finite connected subgraph $G \subset \Gamma$ satisfying the growth condition in Eq. (7), one has $D_G \le |G| - 1$ and $\mathrm{vol}(G) \le C_d|G|$. Combined with the inequalities in Eqs. (7) and (9), this results in the following lower bound for $E_1(G)$:

$$E_1(G) \ge c_d |G|^2. \tag{9}$$

The upper bound by $2h(G)$ is not quite explicit in general (this depends of course on the specific problems at hand), but for finite connected subgraphs $G \subset \Gamma$ one can prove that $\lim_{|G| \to +\infty} \mu_1(G) = 0$ (cf. [14]). Now we are ready to prove the following result.

Lemma 1. Let G be a finite connected subgraph of a graph Γ satisfying the growth condition in Eq. (6), and $0 < \eta \le \frac{1}{6}\lambda_1(-\Delta_G)$. Let V be a nonnegative real function on G, and set $V_\eta(x) := \min[V(x), 2\eta], H_G = -\Delta_G + V$. Then $E_0(H_G) \ge \frac{1}{2}|G|^{-1}\sum_{x \in G} V_\eta(x)$.

Proof. Consider the normalized eigenfunctions of Δ_G with the eigenvalue E_0, viz. $\psi_0 = |G|^{-1/2}\mathbf{1}_G$. Next, introduce an auxiliary operator $K = -\Delta_G + V_\eta$. By nonnegativity of the functions $V_\eta \le V$, we have the inequalities (in the sense of the associated quadratic forms)

$$-\Delta_G \le K \le -\Delta_G + V = H_G \tag{10}$$

so that by the min-max principle, we have $E_0(H_G) \ge E_0(K)$ and $E_1(H_G) \ge E_1(K) \ge E_1(-\Delta_G)$. Since $E_0 = 0$, it follows that $\Delta_G \psi_0 = 0$, and therefore,

$$\begin{aligned} \langle \psi_0, K\psi_0 \rangle &= \langle \psi_0, -\Delta_G \psi_0 \rangle + \langle \psi_0, V_\eta \psi_0 \rangle = \langle \psi_0, V_\eta \psi_0 \rangle \\ &= |G|^{-1} \sum_{x \in G} V_\eta(x) \le 2\eta \le \frac{1}{3}\lambda_1(-\Delta_G) \le \frac{1}{3}\lambda_1(K) \end{aligned} \tag{11}$$

Thus, we have Temple's inequality to ψ_0, K and $E_1(K)$. Note first that by $\Delta_G \psi_0 = 0$, one has

$$\begin{aligned} \langle \psi_0, K^2 \psi_0 \rangle &= \langle \psi_0, (\Delta_G)^2 \psi_0 \rangle + \langle \Delta_G \psi_0, V_\eta \psi_0 \rangle + \langle V_\eta \psi_0, \Delta_G \psi_0 \rangle + \langle \Delta_G \psi_0, (V_\eta)^2 \psi_0 \rangle \\ &= \langle \Delta_G \psi_0, (V_\eta)^2 \psi_0 \rangle. \end{aligned}$$

Now apply Temple's inequality, taking account of the above identity:

$$E_0(H_G) \geq E_0(K) \geq \langle \psi_0, K\psi_0 \rangle - \frac{\langle \psi_0, (V_\eta)^2 \psi_0 \rangle}{E_1(K) - \langle \psi_0, K\psi_0 \rangle}$$

$$\geq |G|^{-1} \sum_{x \in G} V_\eta(x) - \frac{1}{\left(1 - \frac{1}{3}\right) E_1(K)} |G|^{-1} \sum_{x \in G} \left(V_\eta(x)\right)^2$$

$$\geq |G|^{-1} \sum_{x \in G} V_\eta(x) - |G|^{-1} \sum_{x \in G} V_\eta(x) \frac{\frac{1}{3} E_1(K)}{\frac{2}{3} E_1(K)}$$

$$\geq \frac{1}{2} |G|^{-1} \sum_{x \in G} V_\eta(x).$$

□

Lemma 2. Consider a nonnegative i.i.d. random field $V(x, \omega)$ on a locally finite graph Γ satisfying the growth condition in Eq. (7), with the common marginal probability distribution function $F(t) := \mathbf{P}\{V(x, \omega) \leq t\}$, and assume the following:

1. There exist arbitrarily large $L \in N$ such that any ball $B(L, x) \subset \Gamma$ can be partitioned into connected graphs G_i with $L^{\frac{\delta}{12}} \leq |G_i| \leq L^{1/12}$, for some $\delta \in (0, 1)$.
2. $F(t+s) - F(t) \leq Cs^\beta$ for some $\beta \in$, $C > 0$ and all $t \in \mathbb{R}, s > 0$;
3. $\inf\{t \in \mathbb{R} : F(t) > 0\} = 0$

Then for any positive integer n, there exists a finite connected subgraph $G \subset \Gamma$ with $G \vee \geq n$ and satisfying the following spectral bound:

$$\mathbf{P}\left\{E_0(H_G) \leq |G|^{-3}\right\} \leq \exp\left(-\frac{1}{8}|G|\right) \tag{12}$$

Proof. Using the Cheeger bound for the first nonzero eigenvalue, we have

$$E_1(-\Delta_G) \geq c_d |G|^{-2}, c_d > 0. \tag{13}$$

Let $\eta_G = \frac{c_d}{6}|G|^{-2}$, then $2\eta_G \leq \frac{1}{3}E_1(-\Delta_G)$, hence we can use $\eta = \eta_G$ and get

$$E_0(H_G) \geq \frac{1}{2} |G|^{-1} \sum_{x \in G} V_\eta(x, \omega). \tag{14}$$

The value η_G can be made arbitrarily small by taking the cardinality of the graph G large enough, and by assumption on continuity of the probability distribution function F_V, for η_G sufficiently small we have $F_V(\eta_G) \leq 1/4$. Recall that the values of the random potential $V(x, \omega)$ are i.i.d., and so are $V_\eta(x, \omega)$, since they are functions of i.i.d. r.v., so the probability for the sample mean of a large number of i.i.d. random variables to take a value away from the expectation can be assessed with the help of the large deviations theory. Specifically, for any $n \geq 1$ and i.i.d. r.v. $\xi_1, \ldots, \xi_n$, for any $\eta > 0$ such that $\mathbf{P}\{\xi_1 \leq 2\eta\} \leq 1/4$, one has

$$\mathbf{P}\left\{\frac{1}{n}\sum_{i=1}^{n} \min[\xi_i, 2\eta] \le \eta\right\} \le \exp\left(-\frac{n}{8}\right) \tag{15}$$

(see the details in Ref. [11]). Furthermore, we have $|G|^{-3} \le \frac{c_d}{6}|G|^{-2}$ for $|G|$ large enough. Now the lower bound in Eq. (13) combined with Eq. (18) proves the claim:

$$\mathbf{P}\left\{E_0(H_G) \le |G|^{-3}\right\} \le \mathbf{P}\left\{E_0(H_G) \le \frac{c_d}{6}|G|^{-2}\right\} = \mathbf{P}\{E_0(H_G) \le \eta_G\}$$
$$\mathbf{P}\left\{\frac{1}{|G|}\sum_{x \in G} \min[V(x,\omega), 2\eta_G] \le \eta_G\right\} \le \exp\left(-\frac{1}{8}|G|\right). \quad \square$$

Theorem 1. Assume (*W*). There exist some $\delta > 0$ and arbitrarily large balls $B(L,x)$ such that

$$\mathbf{P}\left\{E_0\left(H_{B(L,x)}\right) \le L^{-1/4}\right\} \le \exp\left(\frac{-1}{16} L^{\delta/12}\right). \tag{16}$$

Proof. Fix a vertex $x \in \Gamma$. By assumption (i), there are arbitrarily large L such that the ball $B(L,x)$ can be partitioned into connected graphs G_i with $L^{\frac{\delta}{12}} \le |G_i| \le M(L,x) \le L^{1/12}$. The operator $(-\Delta_G)$ admits the following lower bound in the sense of quadratic forms:

$$-\Delta_G \ge \oplus_{i=1}^{M(L,x)} (-\Delta_{G_i}). \tag{17}$$

Since V is a multiplication operator, we also have

$$H_G \ge \oplus_{i=1}^{M(L,x)} H_{G_i}. \tag{18}$$

Observe that by (i), $L^{-1/4} \le |G_i|^{-3} \le L^{-\delta/4}$. Owing to Lemma 1, we conclude that

$$\begin{aligned}
\mathbf{P}\left\{E_0\left(H_{B(L,x)}\right) \le L^{\frac{-1}{4}}\right\} &\le \mathbf{P}\left\{\min_i E_0(H_{G_i}) \le L^{\frac{-1}{4}}\right\} \\
&\le M(L,x) \max_i \mathbf{P}\left\{E_0(H_{G_i}) \le |G_i|^{-3}\right\} \\
\le C(d)L^d \quad \max_i \exp\left(-\frac{1}{8}|G_i|\right) &\le C(d)L^d \exp\left(-\frac{1}{8}L^{\frac{\delta}{12}}\right) \\
&\le \exp\left(-\frac{1}{16}L^{\delta/12}\right),
\end{aligned}$$

provided that L is sufficiently large. □

Now we give an application of the above result, which was the main motivation in Ref. [11], and explain how the bounds for the eigenvalues of the graph Laplacians give rise to the decay of eigenfunctions. Due to the size limitations of the present chapter, we treat the Green's functions in finite cubes, but the decay of the latter is important in itself, for physical applications, and it is known to imply the decay of eigenfunctions (cf. [10]). The decay of the Green's functions is established by the so-called multiscale analysis (MSA), an inductive scaling algorithm which we will sketch now (details can be found in [9, 10]). First, fix some notations and give some definitions. Given a ball $B = B(L, u) \subset \Gamma$ and the operator $H_B = -\Delta_B + V$, we denote by $G_B(E)$ its resolvent operator H_B, and by $G_B(x, y, E)$, $x, y \in B$, the matrix elements of the resolvent (usually called Green functions) in the standard orthonormal delta-basis formed by the single-site indicator functions 1_x. Further, given a ball $B \subset \Lambda$ inside a larger connected subgraph $\Lambda \subset \Gamma$, the Green functions satisfy an inequality, often called Simon-Lieb inequality (sli), easily following from the second resolvent identity: for any $x \in B$ and $y \in \Lambda \setminus B$, one has

$$|G_\Lambda(x, y, E)| \le \sum_{(v, v') \in \partial B} |G_B(x, v, E)||G_\Lambda(v', y, E)| \tag{19}$$

Since $|B(L, u)| \le C_d L^d$, and the coordination numbers are uniformly bounded by $C_d 1^d = C_d$, we have $|\partial B| \le C_d^2 L^d$, so the above GRI implies

$$|G_\Lambda(x, y, E)| \le C_d^2 L^d \max_{v \in \partial^- B} \quad |G_B(u, v, E)| \max_{v' \in \partial^+ B} |G_\Lambda(v', y, E)| \tag{20}$$

Here, x is an arbitrary point of $B = B(L, u)$, but we will be mostly interested in the case where $x = u$, so the first maximum in the above RHS becomes a characteristic of the ball B.

A simple but important observation is that when $q := C_d^2 L^d \max_{v \in \partial^- B} |G_B(u, v, E)| < 1$, we have for the function $f : x \mapsto |G_\Lambda(x, y, E)|$ a subharmonic-type inequality:

$$0 \le f(x) \le q \max_{v' : d(x, v') = L+1} f(v'), 0 < q < 1. \tag{21}$$

As long as all points v' at distance $L + 1$ from x are centers of L-balls with the same "subharmonic" property, the GRI can be iterated. If n steps of iteration can be performed, and $\|f\|_\infty \le M$ for some $M < \infty$, then the value $f(x)$ admits a small upper bound by Mq^n.

Definition 3. Given real numbers E and $m > 0$, a ball $B = B(L, x)$ is called (E, m)-nonsingular ((E, m)-, NS, in short), if for all $y \in \partial^- B$

$$C_d^2 L^d \max_{v \in \partial^- B} |G_B(u, v, E)| \le exp\Big(- a(m, L)L\Big) \tag{22}$$

with $a(m, L) := m(1 + L^{\frac{-1}{8}})$. Otherwise, it is called (E, m)-singular ((E, m)-S).

The main result of Ref. [11] is the following

Theorem 2. There exist $\delta > 0$, an interval $I = [0, E_*]$ with $E_* > 0$, and an integer L_* such that for all $E \in I$ and $L \ge L_*$ one has

$$\mathbf{P}\{B(L, x) is (E, m) - S\} \le e^{-L^\delta}.$$

We shall need a positive number $\beta \in (0, 1)$; t suffices to set $\beta = 1/2$, which will be assumed below, but for clarity, sometimes the parameter β will be used in its symbolic form.

We denote by $Sp(H_{B,V})$ the spectrum of the operator $H_{B,V}$.

Definition 4. Given an operator $H_{B,V} = -\Delta_B + V$ in a ball $B = B(L, x)$, this ball is called E-nonresonant (E-NR, in short), if $\text{dist}\left(Sp(H_{B,V})\right), E \geq e^{-L^\beta}$, and E-resonant (E-R), otherwise.

Clearly, if a ball is E-NR, the resolvent is well-defined at the energy E, and the modulus of all the respective Green functions are upper bounded by e^{L^β}, since the finite-dimensional operator $H_{B,V}$ is self-adjoint. Probabilistic bounds on $\text{dist}\left(Sp(H_{B,V})\right), E$ for random operators are traditionally called Wegner bounds, due to the original work by Wegner [15] who established the first general bound of that kind.

Lemma 3 (Wegner estimate). Assume that the random potential of the operator $H_{B,V}(\omega) = -\Delta_B + V(\omega)$ is i.i.d. and the common marginal probability distribution of $V(x, \omega)$ admits a probability density bounded by some $C_W < \infty$. Then for any $s \in [0, 1]$

$$\mathbf{P}\{\text{dist}\left(\text{Sp}(H_{B,V})\right), E) \leq s\} \leq C_W |B| s. \tag{23}$$

Definition 5. Given an operator $H_{B,V} = -\Delta_B + V$ in a ball $B = B(L_{k+1}, x)$, this ball is called (E, m)-bad if it contains at least two nonoverlapping (E, m)-S balls of radius L_k, and (E, m)-good, otherwise.

Lemma 4. If a ball $B(L_{k+1}, x)$ is E-NR and (E, m)-good, then it is (E, m)-NS.

Sketch of the proof. The claim is easily obtained by iterating the Simon-Lieb inequality (SLI) and using the hypothesis of (E, m)-goodness; the latter guarantees that in the course of iterated applications of the GRI, one can stumble on an (E, m)-S ball of size L_k at most once. There may be no singular ball inside $B(L_{k+1}, x)$, then the subharmonic-type inequalities easily provide an exponential decay from the center to the boundary of the ball. Furthermore, if there is one singular L_k ball, one can approach it from the center and from the boundary, using the SLI on the first or on the second spatial argument of the Green function. The "wasted" distance is of order or $O(L_k)$, so an elementary calculation provides the desired decay bound of the Green's functions. Technical details can be found in Ref. [16], but it is worth emphasizing the crucial role of the "nonresonance" hypothesis: as was explained, an iterated use of the subharmonic-type inequality in Eq. (21) only gives the upper bound $f(x) \leq q^n \|f\|_\infty$, which is absolutely useless without an explicit control of the sup-norm of the function f, and in our case, one has the functions $f : x \mapsto |G_\Lambda(x, y, E)|$.

Theorem 2 can be derived from the following inductive statement.

Theorem 3. Introduce the following notations: for each $k \geq 0$, let

$$P_k = \sup_{x \in \Gamma} \quad \mathbf{P}\{B(L_k, x) is (E, m) - S\}, \tag{24}$$

$$Q_k = \sup_{x \in \Gamma} \quad \mathbf{P}\{B(L_{k,x}) \quad \text{is} \quad E - R\}. \tag{25}$$

Assume that $P_0, Q_0 \leq e^{-L^\delta}$ with $0 < \delta < \beta = 1/2$, and

$$(YL_0)^\delta \geq 4ln\left(2C_d(2YL_0)^d\right), \tag{26}$$

Then for all $k \geq 0$ one has $P_k \leq e^{-L_k{}^\delta}$.

Sketch of the proof. We proceed by induction, starting with the hypothesis $P_0 \leq e^{-L_k{}^\delta}$. Assume the required bound holds for some $k \geq 0$, then we have to prove it for the balls of size L_{k+1}. By Lemma 4, if a ball $B(L_{k+1}, x)$ is (E, m)-S, then either it is E-R, or it is not (E, m)-good, i.e., contains at least two disjoint balls $B(L_k, u')$, $B(L_k, u'')$ which are (E, m)-S. The number of possible pairs (u', u'') inside $B(L_{k+1}, x)$ is bounded by $\frac{1}{2}C_d^2L_{k+1}^2$, and the probability for each pair $B(L_k, u'), B(L_k, u'')$ to be (E, m)-S is bounded inductively by $P_k \leq e^{-L_k{}^\delta}$. Thus, the probability of existence of at least one such pair is upper-bounded by

$$\frac{1}{2}C_d^2L_{k+1}^2P_k^2 \leq \frac{1}{2}C_d^2L_{k+1}^2e^{-2L_k^\delta}.$$

Further, the probability for the ball $B(L_{k+1}, x)$ to be E-R is upper-bounded with the help of the Wegner estimate from Lemma 111, without induction:

$$Q_{k+1} \leq C_W C_d^2 L_{k+1}^d e^{-L_{k+1}^\beta}, \quad \text{where } \beta = \frac{1}{2} > \delta.$$

Therefore,

$$P_{k+1} \leq \frac{1}{2}C_d^2L_{k+1}^2e^{-2L_k^\delta} + C_W C_d^2 L_{k+1}^d e^{-L_{k+1}^\beta} \tag{27}$$

Now the claim follows by a straightforward, albeit somewhat cumbersome calculation, making use of the assumed geometrical condition in Eq. (29). The details can be found in Ref. [16].

Theorem 3 shows that if on some scale L_0 the Green functions in the balls of radius L_0 decay—with a sufficiently high probability—exponentially fast from the center to the boundary of the ball, then the same phenomenon is reproduced, with ever higher probability, on any scale $L_k \to +\infty$. Such a decay is akin to that of a wave function of a quantum particle in a classically prohibited space where the energy of the particle is below the potential "barrier," so there is a powerful mechanism, originally discovered by P. W. Anderson in 1958, which reproduces the local tendency of a quantum particle to localization in the disordered environment on any scale. The main problem concerns the mechanisms creating such a tendency for localization. This is where we turn to the spectral analysis in the balls $B(L_0, x)$ and seek estimates for the first nonzero eigenvalue of the graph Laplacian in $B(L_0, x)$.

Indeed, if we restrict our analysis to the interval of small positive energies (assuming the potential is nonnegative; otherwise we can make a spectral shift), then it is clear that for all energies E below the spectrum of the operator $G_{B,V}(x, y, E)$ must decay exponentially with respect to the distance $|x - y|$, due to the above-mentioned "under-the-barrier" decay well-known from the elementary exercises in quantum mechanics. Mathematically, there is actually a more general result, the Combes-Thomas estimate [17] which applies not only to the values E

strictly *below* the spectrum of $H_{B,V}$, but all E in the resolvent set, i.e., simply away from the spectrum. Specifically, fix an operator $H_{B,V}$ relative to a ball $B(L,u)$, and let $E \in \mathbb{R}$ satisfy $\text{dist}\left(E, Sp(H_{B,V}) \geq \eta\right)$. There exist universal constants $C, C' > 0$ such that for all $x, y \in B(L,u)$ it holds that

$$|G_{B,V}(x,y,E)| \leq \frac{C'}{\eta} \exp(-C\eta|x-y|). \tag{28}$$

Now all the pieces of the puzzle find their place:

- Using the isoperimetric spectral estimates combined with the Temple inequality, we can find a sufficiently small energy interval $[E, E_* + \eta]$, with $E_*, \eta > 0$, such that in large balls $B(L_0, u)$ a random potential takes a very low average value with a very small probability, so that it is highly unlikely for $H_{B,V}(\omega)$ to have its lowest eigenvalue below $E_* + \eta$.
- Restrict the energy interval to $I_* := [E, E_*]$; then for all $E \in I_*$ we have $\text{dist}\left(E, Sp(H_{B,V}) \geq \eta\right)$, so the Combes-Thomas estimate in Eq. (31) applies and guarantees a fast decay of the Green functions from the center to the boundary of the ball $B(L,)$. Notice that we can have η lower bounded by a fractional power of L. Indeed, Eq. (19) allows us to take $\eta = L^{-1/4}$, and the probability for such a bound to hold is at least $1 - e^{-\delta/12}$, in notations of Eq. (19).
- We thus have, in a tiny interval of energies close to the bottom of the spectrum, the starting hypothesis of the scale induction fulfilled. Now the roll the induction and prove exponential decay with high probability at any scale L_k, $k \geq 0$.

Once again, it is to be stressed that it is a graph-theoretic spectral estimate that makes this story possible, and the presented phenomena take place for a rich class of graphs, much larger than just periodic lattices. This general estimate is a far-going replacement for the elementary consequences of the Fourier analysis on Z^d.

Summarizing, the problem of computing exact asymptotics, or at least sharp upper/lower bounds on the limiting distribution function of the eigenvalues for the operators $H_{B,V}(\omega)$ on various classes of graphs is of course much more general and important than one particular application to the Anderson localization presented above. This if often a difficult problem, and the wealth of knowledge and intuition accumulated in the spectral graph theory would be very welcome to this area of mathematical physics.

3. Symmetric powers of graphs and spectra of fermionic systems

3.1. Motivation and preliminaries

Now we turn to another problem of spectral analysis of quantum Hamiltonians of disordered systems. The presentation will be less technical, and the main message is that the graph theory provides here both an adequate language and technical tools allowing one to treat efficiently difficult problems arising in the recently developed multiparticle localization theory; some of these problems are still open and challenging.

In quantum mechanics, stationary states of a system of several quantum particles are described by the eigenfunctions of their respective Hamiltonians acting in subspaces of either symmetric or antisymmetric functions $\boldsymbol{\Psi}(\mathbf{x})$, $\boldsymbol{x} = (x_1, \ldots, x_N)$, where $N \geq 1$ is the number of particles, and each argument x_j runs through the single-particle configuration space. The particles described by antisymmetric functions are called fermions, and those described by symmetric functions are called bosons. Physically speaking, the particles evolve in the three-dimensional space, but in the framework of the so-called tight-binding approximation, they can be restricted to a periodic lattice or, more generally, a locally finite graph embedded in the Euclidean space. In this section, we assume the latter and work with N-particle systems on a graph Γ. In fact, even the case where $\Gamma = \mathbb{Z}^d$ is of interest for us, since we are going now to show how a typical construction of the graph theory, the symmetric power of a graph, can be instrumental for solving a formal yet thorny technical problem encountered in the multiparticle Anderson localization theory.

The quantum particles are physically indistinguishable, so any accurate mathematical model has to reflect this fact. In some situations including the localization analysis of randomly disordered media, it is more convenient to represent the Hilbert space of symmetric or anti-symmetric functions on Γ as the space of functions on the set of configurations of N indistinguishable particles, instead of a subspace of (+/−)-symmetric functions defined directly on Γ. While the two approaches are mathematically equivalent, the latter one has an important technical advantage that can be explained as follows.

Consider for simplicity of a two-particle fermionic system in a finite subgraph $G \subset [0, L] \subset \mathbb{Z}$ with the graph structure inherited from $\mathbb{Z}$. The wave functions of the two-particle systems are thus antisymmetric functions $\boldsymbol{\Psi}(\mathbf{x}) = \boldsymbol{\Psi}(x_1, x_2)$ of two variables $x_1, x_2 \in G$. We assume the Hamitlonian of this system to ba a discrete Schrödinger operator of the form

Definition 6. *Let be given a random potential $V(\omega)$ on a subgraph G of the lattice Z, and a nonrandom function $r \mapsto U^{(2)}(r)$ of an integer argument. A two-particle discrete Schrödinger operator on G is the operator of the form*

$$\boldsymbol{H}(\omega) = \epsilon \boldsymbol{H}_0 + \boldsymbol{V}(\boldsymbol{x}, \omega) + \boldsymbol{U}(\boldsymbol{x}), \tag{29}$$

where:

- $\boldsymbol{H}_0 = -\Delta_G^{(1)} - \Delta_G^{(1)}$ *(the kinetic energy operator) is the sum of two replicas $\Delta_G^{(j)}$ of the graph Laplacian on G, acting on a function $\boldsymbol{\Psi}(x_1, x_2)$ as a function of the variable x_j, $j = 1, 2$;*
- $\boldsymbol{V}(x, \omega)$ *is the operator of multiplication by the random function $(x_1, x_2) \mapsto V(x_1, \omega) + V(x_2, \omega)$; and*
- $\boldsymbol{U}(x)$ *is the operator of the interaction energy of the two particles at hand, acting as the operator of multiplication by the nonrandom function $(x_1, x_2) \mapsto U^{(2)}(|x_1 - x_2|)$.*

The factor $\epsilon > 0$ measures the amplitude of the kinetic energy operators and reflects the mobility of the particles. In this section, it is instructive to think of ϵ as a small number, so that the potential energy is in a certain sense dominant. $\boldsymbol{V}(\boldsymbol{x}, \omega)$ is the operator of random potential energy induced by the disordered media (modeled here by a finite linear chain) acting as operator of multiplication by the real-valued function

$$(x_1, x_2, \omega) \mapsto \sum_{j=1}^{N} V(x_j, \omega), \quad N = 2,$$

and $V(x, \omega)$ are i.i.d. random variables on G (local potentials produced, e.g., by heavy ions).

The function $U^{(2)}(r)$ is called the two-body interaction potential.

For the sake of notational clarity, here and below we use boldface notations for various objects related to multiparticle objects.

The onset of Anderson localization manifests itself by a fast (usually exponential) decay of the eigenstates $\mathbf{\Psi}_k$ of the Hamiltonian $\mathbf{H}(\omega)$ away from some vertex, depending upon the quantum number \$j\$ (usually referred to as the *localization center* of the respective eigenstate $\mathbf{\Psi}_k$. The quantum transport, on the other hand, may take place due to the tunneling between distant vertices $\mathbf{x}, \mathbf{y} \in G \times G$ with very close local energies. The latter notion can be ambiguous when the kinetic energy is nonnegligent, but pictorially, under the assumption we made above that ϵ is small, the local energy at a vertex x is essentially given by $\mathbf{V}(\mathbf{x}, \omega) + \mathbf{U}(\mathbf{x})$, thus depends directly upon $\mathbf{x}$.

Now recall that the modulus of an asymmetric function $\mathbf{\Psi}(x_1, x_2)$ is symmetric, thus $|\mathbf{\Psi}(\mathbf{x})|$ necessarily takes identical values at vertices $\mathbf{x} = (x_1, x_2)$ and $\mathbf{S}\mathbf{x} = (x_2, x_1)$ in the space of *ordered* configurations of *distinguishable* particles. The symmetric vertices $\mathbf{x}$ and $\mathbf{S}\mathbf{x}$ can be located at arbitrarily large distances from each other in the original two-particle configuration space given by the Cartesian product $G \times G$. As a result, one has, formally, consider the possibility of "tunneling" between $\mathbf{x}$ and $\mathbf{S}\mathbf{x}$, although there is no physical particle transfer process between these two configurations: from the consistent quantum mechanical point of view, the latter are simply identical! We come therefore to realize that the mathematical model based on the Cartesian square of the "physical," single-particle configuration space G generates some formal problems which actually have no physical *raison d'être,* yet they have to be addressed explicitly to rule out some unwanted phenomena. In particular, this renders substantially more complicated the rigorous localization analysis.

However, the above-mentioned difficulty disappears as soon as one replaces the Hilbert space of antisymmetric functions $\mathbf{\Psi}(x_1, x_2)$ on $G \times G$ by an isomorphic Hilbert space of functions $\mathbf{\Phi}$ on the set of configurations of indistinguishable pairs of vertices from the basic graph G. The required construction is well-known in the graph theory: we need a symmetric power $G^{(2)}$ of the graph G. Due to the mathematical complexity of the rigorous multiparticle Anderson localization theory, we can only sketch its general strategy in this chapter, but the main tool, which proves very valuable here, deserves a more detailed discussion. As was said in the introductory part, the main goal of this section is to attract the readers' attention to some interesting and useful relations between the mathematical physics of disordered media and the notions, tools, and deep results of the graph theory.

3.2. Construction of a symmetric power of a graph

3.2.1. *An example in one dimension*

Before we turn to general constructions of symmetric powers of locally finite graphs, it seems instructive to consider first a particular case where the underlying, basic graph G is linear, i.e., isomorphic to a subgraph of the one-dimensional lattice Z. The existence of a complete linear order makes possible a particularly simple variant of the symmetric square $G^{(2)}$ (indeed, of any symmetric power $G^{(N)}$, $N > 1$).

Consider the triangular subset of lattice square $Q(L) := G \times G = [\![0, L]\!] \times [\![0, L]\!]$ (here $G = [\![0, L]\!]$ stands for the integer interval $[0, L] \cap \mathbb{Z}$):

$$G^{(2)} = \{(x_1, x_2) \in [\![0, L]\!] \times [\![0, L]\!] : x_1 < x_2\}$$

(here (2) reflects the number of "particles"). An example is presented in **Figure 1**. Any anti-symmetric function $\mathbf{\Psi}$ on $Q(L)$ vanishes at any point of the form (x, x), and its modulus takes identical values on $(x_1, x_2) \in G^{(2)}$ and on the symmetric point (x_2, x_1). It follows that

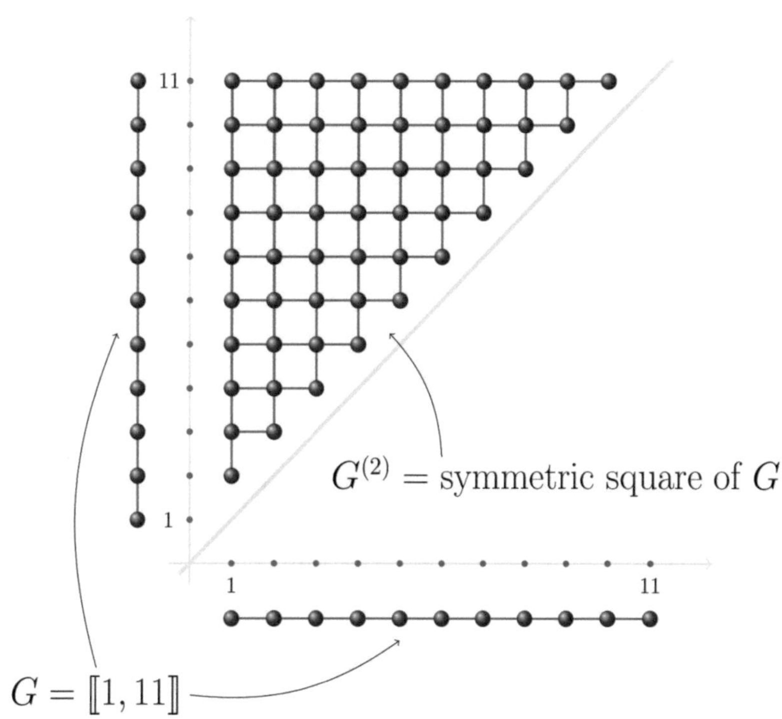

Figure 1. Example of a symmetric square. Here the base graph G is a subgraph of Z, and it can be implemented as a subgraph of the Cartesian square Z^2, owing to the one-dimensional topology of Z.

$$\|\Psi\|^2_{2,Q(L)} := \sum_{x\in Q(L)} |\Psi(x)|^2 = 2\sum_{x\in A^{(2)}} |\Psi(x)|^2 \tag{30}$$

This provides a natural isomorphism between the Hilbert spaces of complex functions on $Q(L)$ and on $G^{(2)}$. Clearly, the same idea works in the N-particle case, where we can define

$$G^{(N)} = \{(x_1, \ldots, x_N) \in [\![0, L]\!]^N : x_1 < x_2 \ldots < x_N\},$$

except that in the latter case, the factor of $2 = 2!$ in Eq. (23) is to be replaced by $N!$.

Such a simple, transparent geometrical construction is no longer available for general graphs, but an isomorphism similar to that from Eq. (23) can be established for the symmetric powers of graphs. Below we give a variant with a distinctive flavor of quantum mechanics.

3.2.2. General construction

The vertex set. Let be given a connected, locally finite countable graph with the vertex set G and an edge set E, and an integer $N \geq 2$. Consider the integer-valued functions $\mathbf{n} : x \mapsto \mathbf{n}(x) \in \{0, 1\}$ on G such that $\sum_{x\in G} \mathbf{n}(x) = N$. The physical meaning of the value $\mathbf{n}(x)$ is the number of particles at the vertex x, so it is usually called the occupation number of the site $x \in G$. Due to the indistinguishable nature of the particles, only the numbers of particles at each site are physically observable (measurable in experiments). Furthermore, since we are modeling now fermions, the respective wave functions, by their antisymmetry, must vanish on any configuration of N particles among which at least two occupy the same position. This was precisely the reason we excluded the "diagonal" from $G^{(2)}$ above. Now we achieve the same effect by the requirement $\mathbf{n}(x) \in \{0, 1\}$. The bottom line is that a configuration of N particles admissible for modeling fermions is completely determined by a function $\mathbf{n}$; for all intents and purposes, each $\mathbf{n}$ is a (fermionic) configuration.

Hence, we constructed an appropriate vertex set $G^{(N)}$ of the graph that would generalise $G^{(2)}$ for an general underlying graph G. Specifically, there is a projection

$$\Pi^{(N)} : \mathbf{x} = (x_1, \ldots, x_N) \mapsto \mathbf{n}_x, \text{ where } supp\ \mathbf{n}_x = \{x_1, \ldots, x_N\}. \tag{31}$$

The points of the support of a function $\mathbf{n}_x$ will be called the *particles of the configuration* $\mathbf{n}_x$. Restricted on the set of N-tuples of pairwise distinct vertices $x_1, \ldots, x_N$, the projection $\Pi^{(N)}$ is exactly $N!$-fold: the pre-image $\left(\Pi^{(N)}\right)^{-1}(\mathbf{n})$ has cardinality $N!$.

The edge set. Using again a terminology inspired by physics, we say that two configurations $\mathbf{n}', \mathbf{n}''$ form an (unordered) edge if and only if $\mathbf{n}''$ is obtained by moving exactly one particle of $\mathbf{n}'$ to a position unoccupied by other particles from $\mathbf{n}'$.

Mathematically, we require that

http://dx.doi.org/10.5772/intechopen.68480

$$\sum_{x\in G}|\mathbf{n}'(x) - \mathbf{n}''(x)| = 2; \tag{32}$$

It is not difficult to see that the two definitions are equivalent. Indeed, each term in the above sum equals 0 or 1, since $0 \le \mathbf{n}'(x), \mathbf{n}''(x) \le 1$, and such a term vanishes when either $\mathbf{n}'(x) = \mathbf{n}''(x) = 0$, i.e., x is unoccupied by either configuration, or $\mathbf{n}'(x) = \mathbf{n}''(x) = 1$, i.e., both configurations have a particle at x. Removing particles from the support of $\mathbf{n}'$ to produce new configuration $\mathbf{n}''$, we have to place them outside the support. Therefore,

each point x with $\mathbf{n}'(x) = 1$ and $\mathbf{n}''(x) = 0$ (i.e., occupied by $\mathbf{n}'$ but unoccupied by $\mathbf{n}''$) contributes by a two unit term to the sum: first, $\mathbf{n}'(x) - \mathbf{n}''(x) = 1$, and second, for the position y to which we move the particle from x we have $\mathbf{n}'(x) - \mathbf{n}''(x) = -1$. If we move more than one particle from $\mathbf{n}'$, the sum in Eq. (25) will be at least 4. We conclude that the second definition in Eq. (25) is equivalent to the first one.

This completes the construction of the N-th symmetric power $(G^{(N)}, E^{(N)})$ of the graph (G, E).

3.2.3. *Hilbert space isomorphism: antisymmetric functions versus symmetric power*

Now the isomorphism between of the Hilbert space of square-summable complex antisymmetric functions on the Cartesian power G^N and the Hilbert space of all square-summable complex functions on $G^{(N)}$ is defined in a natural fashion. Recall that we introduced the projection $\Pi^{(N)}$, such that $\left(\Pi^{(N)}\right)^{-1}(\mathbf{n}_x)$ consists of all $N!$ permutations of the vertices $x_1, \ldots, x_N$ from the support of $\mathbf{n}_x$. Any function $\mathbf{\Phi}$: $G^{(N)} \to C$ generates a symmetric function $\mathbf{\Psi} = \mathbf{\Phi}\circ\Pi^{(N)}$, and each symmetric function $\mathbf{\Psi} : G^N \to C$ can be obtained in this way, as $\mathbf{\Phi}\circ\Pi^{(N)}$. The modulus of any antisymmetric function $\mathbf{\Psi}$ on the Cartesian power G^N takes identical values on all elements of $\left(\Pi^{(N)}\right)^{-1}(\mathbf{n}_x)$. Thus, with $\mathbf{\Psi_\Phi} := \mathbf{\Phi}\circ\Pi^{(N)}$,

$$\sum_{x\in G^N}|\mathbf{\Psi_\Phi}(x)|^2 = N!\sum_{\mathbf{n}_{x\in G^{(N)}}}|\mathbf{\Phi}(x)|^2. \tag{33}$$

Now the required isomorphism is defined by $\|\mathbf{\Psi_\Phi}\|_2 = \sqrt{N!}\|\mathbf{\Phi}\|_2$.

It might appear that the described isomorphism is just an interesting formal trick, but there is much more to it than a mere mathematical curiosity. As one illustration, we consider now a problem of spectral analysis for multiparticle random Hamiltonians $\mathbf{H}(\omega)$ above.

3.2.4. *An application: KAM-type analysis of two-particle fermionic random Hamiltonians*

The goal of this subsection is merely to illustrate the key role of the isomorphism between the subspace of antisymmetric square-summable functions of $N > 1$ variables in a graph G and the space of all square-summable functions on the symmetric power $G^{(N)}$. The mathematical problem where it is used is quite complex, so we will only sketch the main setting and focus on the role of the symmetric powers.

It is to be emphasized that the spectral analysis of N-particle quantum Hamiltonians in presence of a random potential field, generated by a disordered media, accurately taking into account a nontrivial interaction between the particles, is a relatively new direction both in theoretical physics and in rigorous mathematical physics. This is an actively developing and challenging area of research. While physicists have finally obtained convincing theoretical results on the stability of localization under the Coulomb interaction, the progress in rigorous mathematical physics is still relatively modest, compared to theoretical physics. The best insight achieved in the last 10 years concerns the systems of a fixed number N of particles in a large sample of a disordered media. For the purposes of this chapter, we always restrict the analysis to discrete systems on graphs.

As was said in the introductory section, we can only sketch rather complex mathematical constructions involved and illustrate the main mechanisms of stability of localization under interaction. For simplicity, suppose that we have a system of $N = 2$ particles in a large but finite connected graph G on which an i.i.d. random (potential) field $x \mapsto V(x, \omega)$ is defined. One can consider various marginal probability distributions, i.e., identical probability distributions of the random variables $V(x, \omega), x \in G$. Two models popular in theoretical physics of disordered media suit perfectly our needs here: a standard Gaussian distribution $N(0, 1)$ with zero mean and unit variance, and the uniform distribution on the interval $[0, 1]$.

Consider first the simplest (yet mathematically nontrivial) case of zero amplitude of interaction. Then the variables in the Hamiltonian $\mathbf{H}(\omega)$ can be separated, since in this case one has an algebraic representation $\mathbf{H}(\omega) = \mathbf{H}^{(1)}(\omega) \otimes \mathbf{I} + \mathbf{I} \otimes H^{(2)}(\omega)$ where I is the identity operator, and therefore, the eigenvectors of the operator $\mathbf{H}(\omega)$ can be chosen in the tensor product form, $\mathbf{\Psi}_{i,j}(\omega) = \psi_i(\omega) \otimes \psi_j(\omega)$, where $\psi_i(\omega), \psi_j(\omega)$ are eigenvectors of the single-particle Hamiltonians $H^{(1)}(\omega)$ and $H^{(2)}(\omega)$, respectively. The latter are identical replicas acting on their respective variables x_1 and x_2. The eigenvalues are the sums $E_{i,j}(\omega) = E_i^{(1)}(\omega) + E_j^{(2)}(\omega)$, with $\mathbf{H}^{(1,2)}(\omega)\psi_i(\omega) = E_i^{(1,2)}(\omega)\psi_i(\omega)$.

The main reason why the electron-electron interaction was consciously neglected even in theoretical physics, is that it is relatively weak as compared to the potential energy of interaction with the ions surrounding the mobile electrons, so we also assume the amplitude of the interaction potential to be small.

Another important assumption can simplify the spectral analysis, at least in an informal treatment of the problem: small amplitude of the factor ϵ in front of the kinetic energy operator $\mathbf{H}_0$; in physical terms, this corresponds to a low mobility of the particles at hand which, naturally, should favor "localization" of a given particle. Mathematically, already for $N = 1$ (isolated particles), with $\epsilon = 0$ we get a multiplication operator which has the orthonormal eigenbasis composed of the "discrete delta-functions" $\phi_x = \mathbf{1}_{\{x\}}, x \in G$. Similarly, for $N \geq 2$ and $\epsilon = 0$ we have a perator of multiplication by the total potential energy which also has a complete eigenbasis composed of "localized" eigenfunctions.

If we had constructed an eigenbasis of the multiplication operator in the representation on the Cartesian square G^2, then we would have obtained the two-site supported eigenfunctions:

http://dx.doi.org/10.5772/intechopen.68480

$$\frac{1}{\sqrt{2}}\mathbf{1}_{\{x\}} \otimes \mathbf{1}_{\{y\}} - \frac{1}{\sqrt{2}}\mathbf{1}_{\{y\}} \otimes \mathbf{1}_{\{x\}}$$

while the representation on the symmetric square $G^{(2)}$ gives rise to single-site eigenfunctions $\mathbf{1}_{\{x\}} \otimes \mathbf{1}_{\{y\}}$, where $x \neq y$. Naïvely, starting with the "ultralocalized" eigenfunctions $\mathbf{1}_{\{x\}} \otimes \mathbf{1}_{\{y\}}$, one may attempt to use the first-order perturbation theory. The well-known formulae of the rigorous perturbation theory for the eigenvectors reveal two problems:

- "small denominators," i.e., pairs of very close or equal eigenvalues; and
- large dimension of the spectral problem, which may also give rise to degenerate eigenvalues or at least to some pairs of close eigenvalues.

Indeed, the eigenvalue associated to the unperturbed eigenfunctions of the potential energy operator $\mathbf{\Phi}_{x,y} = \mathbf{1}_{\{x\}} \otimes \mathbf{1}_{\{y\}}$ is given by

$$E^0_{x,y} = V(x,\omega) + V(y,\omega) + U(|x-y|) \tag{34}$$

Fix now the random field model with uniform marginal distribution, and let the interaction be uniformly bounded, then the above eigenvalues all belong to some fixed, bounded interval, regardless of the dimension $D = G\vee(|G|-1)$ of the Hilbert space, growing with the cardinality of G. The larger is $G\vee$, the closer the D eigenvalues must get, counted with multiplicity and restricted to a fixed interval of $\mathbb{R}$. In the Gaussian model, a similar phenomenon is encountered, with high probability, since large values of the Gaussian random potential $V(x,\omega)$ are taken with small probability.

A conclusion we can draw from this elementary analysis is that one cannot expect the perturbation theory for nondegenerate spectra, or for the finite-dimensional operators with bounded multiplicity, to work efficiently in the model with a large graph G.

Several approaches have been developed in spectral theory of single-particle random Hamiltonians in the last three decades. Technically, they are based on different mechanisms, and even a brief presentation of these approaches would require an entire book. Interested readers may familiarize themselves with the basic techniques in the monographs [9, 10]. Recently, there has been a wake of growing interest to the technique going back to the celebrated KAM (Kolmogorov-Arnold-Moser) theory originally developed for the analysis of stability of invariant tori in some nonlinear dynamical systems

Recently, Imbrie [18] adapted the KAM techniques to the spectral analysis of random lattice Hamiltonians, in any dimension, and to one-dimensional random spin chains.

In essence, the "linear" version of the KAM method is an inductive, iterated use of the first-order perturbation theory with an accurate account of the higher-order terms represented by an infinite number of diagrams. At each step of the induction, one obtains an orthogonal basis for the considered (random) operator $H(\omega)$ that is an approximate eigenbasis for the latter, but with better and better accuracy. The error terms on the k-th step of induction feature a typical Newtonian decay rate like e^{-q^k}, where $q \in (1,2)$, which is not surprising since KAM technique is based on one or another form of Newton's method. Once a new, more accurate eigenbasis of order k is constructed by perturbing its predecessor of order $(k-1)$, the matric elements of $H(\omega)$ are computed in this (orthonormal) basis, and the process is repeated. KAM type constructions are usually quite complex and cumbersome. One has first to describe in detail the entire set of properties to be assumed on the step $k-1\geq 0$ and then reproduced on the next induction step k.

A synthetic method, employing some ideas of the KAM method and other, simpler techniques elaborated in the spectral theory of single-particle random Hamiltonians, have been proposed first to the noninteracting random systems [11], and later on to their N-particle counterparts. The pivot of this method, like in the KAM approach, is an accurate quantitative control of the "small denominators"—minimal distances between distinct random eigenvalues of random Hamiltonians associated with finite but growing subgraphs of a countable graph. It would be difficult to carry out such a program in the representation of distinguishable particles, i.e., on the Cartesian power G^N.

Now return to the semiquantitative analysis of a two-particle Hamiltonian. Restrict ourselves to a case where the size of the underlying graph G, modeling the "physical" space where the two quantum particles evolve, has a fixed size, and allow us to vary the parameter ϵ in $\epsilon \boldsymbol{H}_0$ (the mobility amplitude), and to take it as small as needed for an attempt to make one step of application of the perturbation theory for nondegenerate spectra.

With both ϵ and h small enough, the main contribution comes from the random potential, so we have the eigenfunctions of the unperturbed operator $\boldsymbol{\Phi}_{x,y}\boldsymbol{\Phi}_{x,y} = \mathbf{1}_{\{x\}} \otimes \mathbf{1}_{\{y\}}$ with eigenvalues $\lambda_{x,y} = \boldsymbol{V}(x,\omega) + \boldsymbol{V}(y,\omega)$, and we have to assess the difference between two such eigenvalues, labeled by two pairs of sites (x,y), (x',y') of the graph G:

$$\lambda_{x,y} - \lambda_{x',y'} = \Big(V(x,\omega) + V(y,\omega)\Big) - \Big(V(x',\omega) + V(y',\omega)\Big).$$

Since the potential is random, there can be no uniform, deterministic lower bound on the absolute value of the above difference: with positive probability, it can be smaller than any $\delta > 0$. The randomness, however, is a double-edged sword: while small values of the difference are certainly possible, they may, or might, be unlikely, so we have to determine, how unlikely is to have $|\lambda_{x,y} - \lambda_{x',y'}| < \delta$.

To begin with, notice that we have to deal with different eigenfunctions, hence with two nonidentical pairs (x,y), (x',y'). Thus, $\mathrm{card}\{x,y\} \cap \{x,y\} \leq 1$. Consider two cases.

I. $\mathrm{card}\{x,y\} \cap \{x',y'\} = 0$.

In this case, the random variables $\lambda_{x,y} = V(x,\omega) + V(y,\omega)$ and $\lambda_{x',y'} = V(x',\omega) + V(y',\omega)$ have no common terms, and therefore they are independent. Moreover, inside each pair we have independence, so the probability distribution of each sum can be easily obtained by convolution. For simplicity, consider the case of standard Gaussian variables, then each eigenvalue is also Gaussian with zero mean and variance 2. The difference $\lambda_{x,y} - \lambda_{x',y'}$ is again a sum of two i.i.d. Gaussian variables $\lambda_{x,y}$ and $-(\lambda_{x',y'})$, hence it is Gaussian with zero mean and variance 4. Recalling the explicit form of the Gaussian probability density with variance σ^2, which is uniformly bounded by $1/\sqrt{2\pi\sigma^2}$, we conclude that

$$\mathbf{P}\{|\lambda_{x,y} - \lambda_{x',y'}| < \delta\} \le \frac{\delta}{2\sqrt{2\pi}} \tag{35}$$

II. $\quad \mathrm{card}\{x, y\} \cap \{x', y'\} = 1.$

Now the unordered pairs x, y and x', y' have exactly one common point; let it be x, so we have a pair of eigenvalues $\lambda_{x,y}$ and $\lambda_{x,y'}$ with $y \neq y'$ and the difference given by

$$\lambda_{x,y} - \lambda_{x',y'} = \Big(V(x,\omega) + V(y,\omega)\Big) - \Big(V(x,\omega) + V(y',\omega)\Big) = V(y,\omega) + \Big(-V(y',\omega)\Big),$$

so it is again a sum of two independent random variables. Assuming as before the distribution to be Gaussian with zero mean and unit variance, we see that $\lambda_{x,y} - \lambda_{x',y'}$ is centered Gaussian with variance 2, so the analog of Eq. (32) is now

$$\mathbf{P}\{|\lambda_{x,y} - \lambda_{x',y'}| < \delta\} \le \frac{\delta}{2\sqrt{\pi}} \tag{36}$$

The final conclusion is that for small $\delta > 0$, the small differences between any two eigenvalues corresponding to two distinct eigenfunctions on the symmetric square of the underlying graph is small, viz., of order of $O(\delta)$. Therefore, for a finite graph of size $|G| = n$, the probability to have at least one pair of eigenvalues at distance smaller than δ is bounded by $\frac{\mathbf{n}(\mathbf{n}-1)\delta}{4\sqrt{\pi}} = O(n^2\delta)$ (we used the weakest of the two estimates in Eqs. (32) and (33)).

This simple probabilistic analysis provides the logical basis for the KAM approach, where we can rule out "small denominators" which cannot be tolerated in the analytic application of the first-order perturbation formulae, at some initial scale. The rest of the procedure requires a number of analytic efforts, but the crucial point, viz., the possibility to avoid degenerate eigenvalues at the initial scale, is the direct consequence of the graph-theoretical construction of a symmetric power of a graph G. Using a Cartesian power of G would at best significantly complicate the entire procedure, and perhaps render it impractical. In any case, no replacement for resorting to symmetric powers has been found so far in multiparticle localization theory of fermionic systems on graphs.

4. Combinatorial and metric (quantum) graph

Our final topic also concerns the constructive relations and interactions between the graph theory, in a broad sense, and the mathematical physics of the quantum world. However, the general direction of these interactions will be reversed, for we are going to discuss a very recently developed class of mathematical objects naturally emerged in the analysis of interacting quantum systems. We would like to attract the attention of experts in graph theory and related fields to the new area, where a number of questions are not even properly formulated, and many interesting phenomena are yet to be discovered.

First, recall the notion of a metric graph; due to a wake of interest to the so-called nanotubes, one often refers to these mathematical objects as "quantum graphs."

Metric graphs represent an important link between the discrete spaces and manifolds endowed with a rich local structure of a Euclidean space. By definition, a metric graph $Q = Q_\Gamma$ over a finite or countable unoriented combinatorial graph (Γ, E), with the vertex set Γ and the edge set E, is a singular one-dimensional manifold constructed as follows. Associate with each edge $e = (\iota, \tau) \in E$ an open interval I_e, considered as a Riemannian manifold with the Riemannian metric inherited from $\mathbb{R}$. In some models, all the intervals have the same lengths, so by a change of parameters one usually can assume they are replicas of $(0,1)$. In other models, on the contrary, one allows variable length of these basic intervals. We will assume the former and work with unit intervals. There is a canonical oriented graph associated with the unoriented graph (Γ, E), with the same vertex set and two opposite edges for each edge in E. In some auxiliary constructions, this morphism from the category of unoriented graphs to that of oriented ones can be used, to avoid some ambiguities, but it will be not crucial to our purposes, since we will work with a second-order differential operator (essentially the second derivative operator), so the orientation will not be really important.

Each open interval $I_e \cong (0,1)$ is canonically compactified by its natural embedding into $[0,1]$. Taking an edge (x, y) and fixing its orientation in one of the two possible ways, so that $(x, y) \cong (\iota, \tau)$, we thus can identify its starting point ι with $0 \in [0,1]$ and the terminal point τ with $1 \in [0,1]$. Next, we define the differential operator $L = -d^2/dt^2$ in the space of twice differentiable functions on $(0,1)$; boundary conditions are discussed below. In other words, $L = -\Delta$, where Δ is the Laplacian on the Riemannian manifold $(0,1)$. While d/dt requires a local coordinate, hence a fixed orientation, L is not sensible to this choice.

Further, consider the disjoint union $Q_\Gamma^{(0)}$ of the basic (open) intervals I_e, finite or countable, with the natural structure of the measure space induced by the Lebesgue measure on each interval with the respective sigma-algebra of measurable subsets. In turn, this allows us to introduce the Hilbert space of square-integrable functions on $Q_\Gamma^{(0)}$; this is not yet an object we had needed, for there is no connections between the restrictions of a given function f on $Q_\Gamma^{(0)}$ to different, pairwise disjoint connected components thereof.

Now it is time to choose boundary conditions, having in mind the canonical embedding of $Q_\Gamma^{(0)}$ into the union Q_Γ of the compactified intervals $\bar{I}_e \cong [0,1]$. In application to the "quantum" graphs, traditionally one imposes the Kirchhoff conditions. Now, for formal reasons, fix some orientation on each edge, hence, a local coordinate on each $\bar{I}_e \cong [0,1]$. Then we can define the one-sided first derivatives on each vertex, in the directions of all attached intervals $\bar{I}_e$. Let D_e be such a derivative along the local coordinate on $\bar{I}_e$, and set $c_e = 1$ for outgoing edges and $c_e = -1$ for the ingoing ones. The Kirchhoff conditions are as follows: a function f must be continuous at each vertex and obey a conservation law

$$\sum_e c_e D_e f = 0. \tag{37}$$

Below we call the intervals $\bar{I}_e$ *continuous* edges.

Now, using the standard methods of functional analysis, one can construct a self-adjoint extension of the "Laplacian" L with Kirchhoff boundary conditions, and for any, say, bounded measurable function $V : Q_\Gamma \to R$ (a potential), define the Schrödinger operator H_V as unbounded self-adjoint operator in $H = L^2(Q_\Gamma)$, with the suitable domain.

One can perturb the above, rather idyllic picture in several ways. First, one can consider a random potential $V(\omega)$, taking i.i.d. random values on each edge. Further, on can vary the lengths l_e of the continuous edges, assuming that $l_e(\omega)$ are i.i.d. random variables with a common probability distribution. From the functional analytic point of view, treating unbounded self-adjoint operators on metric graphs, in the framework of random operators, is substantially more delicate a matter that the analysis of finite-difference operators on the underlying discrete, combinatorial graphs. One may wonder, whether some properties of the Hamiltonians on the underlying graph can be useful for the analysis of their continuous siblings Q_Γ. The theory of boundary triples (cf. [19]) provides a powerful and valuable tool of spectral analysis on continuous metric graphs, where an essential part of technical work is carried out in a simpler framework of countable graphs with discrete Schrödinger operators.

Now we turn to a further development in this direction made recently by Sabri [7] who introduced the notion of multiparticle quantum graph. We consider the simplest nontrivial case of $N = 2$ quantum particles on a quantum graph Q_Γ. To be able to refer to existing results and publications, we assume the particles distinguishable.

In the discussion of two-particle systems on a graph G in Section 3, the pair (x_1, x_2) was ranging in the Cartesian (and then symmetric) square of G, and the latter is, topologically, a discrete space, thus essentially of the same nature as the factors in the product $G \times G$. But now that the configuration space Q_Γ for each particle is a continuous object, viz. a (singular) one-dimensional Riemannian manifold, the situation changes radically: the configuration space for the pair (x_1, x_2) is locally a *two*-dimensional manifold; in the case of an N-tuple $(x_1, \ldots, x_N)$ it becomes N-dimensional. Many specifically 1D methods of spectral analysis are inapplicable in dimension $d \geq 2$.

Shortly after the publication of the first results on N-particle Anderson localization in periodic lattices and in Euclidean spaces, Sabri [7] proposed an interesting extension of the new techniques and results to the multiparticle systems on quantum graphs. His construction was essentially motivated by a specific goal, but there are various contexts where the construction itself may prove valuable.

For $N = 2$, one has to start again with the building blocks of a 1D quantum graph Q_Γ over a combinatorial graph Γ: the open finite intervals associated with each edge of Γ. Restricting the positions x_i of the two particles to their respective continuous edges identified with $[0, 1]$, we have the pair (x_1, x_2) ranging in the unit square $[0, 1] \times [0, 1]$. For brevity, call such basic squares *cells*. Each cell is delimited by four continuous edges inherited from the first and from the second particle, and these four (continuous) edges are the loci of contact between the cells. Clearly, this complicates the structure of what was the edge set in the underlying graph Γ, but this is the natural replacement for the notion of the edge set; this is what defines the topology and metric geometry of the new object, called by Sabri a two-particle (more generally, N-particle) quantum graph $Q_\Gamma^{(2)}$.

Just like the Laplacian L defined on the quantum graph, we can define its two-particle counterpart $L^{(2)}$: first, on the unit squares, and then proceed to self-adjoint extensions with one or another kind of boundary conditions to be imposed on the 1D continuous edges of the conventional, one-particle quantum graphs supporting each of the two particles. This inevitable functional analytic work has been done by Sabri. And of course, once the natural Laplacian $L^{(2)}$ is defined as an unbounded self-adjoint operator with a suitable domain in the Hilbert space of square-integrable functions on $Q_\Gamma^{(2)}$, one can also define the Schrödinger operators $\mathbf{H}_V^{(2)} = -L^{(2)} + V$, e.g., for bounded measurable functions V. In **Figure 2**, we give an example of three models based on the same graph structure: a combinatorial graph, a quantum graph, and a two-dimensional domain surrounding the quantum graph in question.

The new construction raises a number of questions, of different nature. One of them concerns the constructive relations between the spectral properties of a Schrödinger operator $\mathbf{H}_V^{(2)}$ on the continuous, locally two-dimensional (2D) manifold $Q_\Gamma^{(2)}$, and its analog on the Cartesian square of the combinatorial graph Γ.

Another question, of functional analytic nature, raised by Sabri, concerns an explicit description of the self-adjoint extensions of the 2D Laplacian initially defined, say, on infinitely differentiable functions with support inside an open cell $\cong (0,1)^2$. It appears that the corner

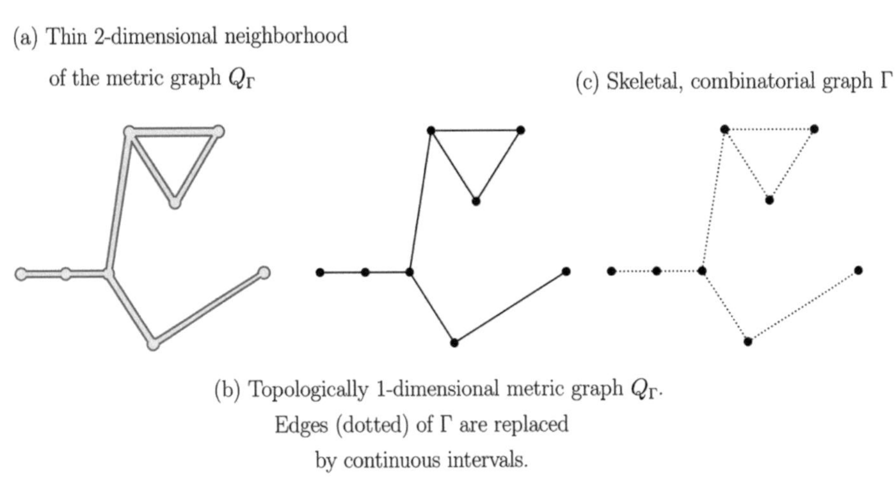

Figure 2. Example of (a) physical, thin two-dimensional area A; (b) corresponding metric graph Q_Γ: a mathematical abstraction where the finite width of A is ignored; and (c) the combinatorial graph with the same vertices as Q_Γ.

points make the explicit analysis difficult, although the existence of the desired extensions poses no serious problem.

5. Conclusion

Mathematical modeling of physical phenomena had provided important motivations for developing various fields of mathematical physics since several centuries; as to the quantum physics, its development was from the beginning of the twentieth century parallel to the development of the functional analysis in general and spectral theory of operators in particular. The remarkable discovery made by P. W. Anderson in 1958 brought to life a synthetic approach to modeling disordered systems based on a fusion of analysis in a broad sense with probability theory. The physical community came to realize that the models based on the idealized picture of perfectly periodic crystals miss some crucial mechanisms responsible for transport (e.g., electrical conductivity) or absence thereof under the Anderson localization. The classical formulae for conductivity and many related phenomena, crucial for the development of modern microelectronics and nanotechnologies, cannot ignore the localization/delocalization problematics. While the most simple models may refer to the integer (and some other periodic) lattices in a Euclidean space where classical Fourier analysis can use the method of separation of variables, the situation can be significantly more complex in the case of quasicrystals, featuring both a local order and long-range disorder. Mathematically, such structures are described as nonperiodic graphs where the Fourier analysis breaks down, and one needs some efficient, constructive replacements. Furthermore, large and complex molecules studied in organic chemistry and molecular biology also require a versatile toolbox not limited to a commutative Fourier analysis. Also, the crystalline media in presence of structural (e.g., mechanical) defects are not *stricto sensu* periodic, so again one needs robust eigenvalue distribution bounds not relying on the exactly periodic geometry of the media. In Section 2, we have seen that the isoperimetric inequalities, appeared in the graph theory under the influence of its diverse applications, provide indeed adequate tools for an asymptotic analysis of the limiting eigenvalue distribution for discrete quantum Hamiltonians used in physics in the framework of the so-called tight-binding approximation effective for the "low" energies. The term "low" actually refers to the energies lost important to the quantum processes exploited in modern microdevices (e.g., CPU having diameter of a few millimeters and width of order of a few dozens of atomic layers), in biological tissues and technologically created organic substances.

In 2008, The Isaac Newton Institute for Mathematical Sciences in Cambridge, Great Britain, has organized a semiannual program "*Mathematics and Physics of Anderson Localization: 50 Years Later*" aiming to summarize the impact of Anderson's theory on physical theories and applications as well as on the mathematical physics. The general conclusion many participants and younger researchers could draw from numerous and diverse presentations was that the paradigm of quantum localization/delocalization provides today both a language and a general theoretic background for many specific directions of research; it is not an isolated pragmatic physical model or abstract mathematical problem. The program in question also revealed to the physics and mathematics communities the importance of the interparticle interaction which was briefly discussed in Section 3; the need for such theory was emphasized already by Anderson in his early papers, but one had to wait almost half a century to see its emergence in independent physical and rigorous mathematical works. Shortly after the program, the Anderson localization theory for interactive disordered systems has been applied (in mathematical works) to the nanotubes modeled by quantum graphs. While the size limitations of the present work do not allows us to present mathematical details of the new theory, there is no doubt that many of its mathematical aspects are closely related to the methods of the graph theory. Further reading, along with an extensive bibliography, can be found in the first monograph [4] dedicated to localization phenomena in interactive systems. This new direction of mathematical physics still is at its early stage of development. The language and toolbox of the graph theory proved to be very useful here, as we have seen in Section 3. On the other hand, new structures discussed in Section 4, emerging from the analysis of multiparticle quantum graphs open new problems and propose new types of models to the graph theory. This chapter was written in the hope to bring closer the communities of researchers, particularly the younger ones, working in functional analysis, graph theory in a broad sense, and in probability theory.

Author details

Victor Chulaevsky

Address all correspondence to: victor.tchoulaevski@univ-reims.fr

Department of Mathematics, University of Reims, Reims Cedex, France

References

[1] Anderson PW. Absence of diffusion in certain random lattices. Physical Review. 1958;**109**(5): 1492-1505

[2] Basko DM, Aleiner IL, Altshuler BL. Metal-insulator transition in a weakly interacting many-electron system with localized single-particle states. Annalen der Physik. 2006;**321**(5): 1126-1205

[3] Chulaevsky V, Suhov Y. Eigenfunctions in a two-particle Anderson tight binding model. Communications in Mathematical Physics. 2009;**289**:701-723

[4] Chulaevsky V, Suhov Y. Multi-particle Anderson localisation: Induction on the number of particles. Mathematical Physics Analysis and Geometry. 2009;**12**:117-139

[5] Aizenman M, Warzel S. Localization bounds for multiparticle systems. Communications in Mathematical Physics. 2009;**290**:903-934

[6] Chulaevsky V, Suhov Y. Multi-scale Analysis for Random Quantum Systems with Interaction. Boston: Birkhäuser; 2013. p. 236

[7] Sabri M. Anderson localization for a multi-particle quantum graph. Reviews in Mathematical Physics. 2014;**26**(1). DOI: 10.1142/S0129055X13500207.

[8] Lifshitz IM, Gredescul SA, Pastur LA. Introduction to the Theory of Disordered Systems. New York: Wiley; 1988. p. 358

[9] Carmona R, Lacroix J. Spectral Theory of Random Schrödinger Operators. Boston, Basel, Berlin: Birkhäuser; 1990. p. 587

[10] Stollmann P. Caught by Disorder. Boston: Birkhäuser; 2001. p. 166

[11] Chulaevsky V. Direct scaling analysis of localization in single-particle quantum systems on graphs with diagonal disorder. Mathematical Physics Analysis and Geometry. 2012;**15**:361-399

[12] Temple G. The theory of Rayleigh's principle as applied to continuous systems. Proceedings of the Royal Society of London Series A. 1928;**19**:276-293

[13] Chung FRK. Spectral Graph Theory. Washington, DC: CBMS Conference Series in Mathematics; 1997. p. 117

[14] Chung FRK, Grigor'yan A, Tau S-T. Upper bounds for eigenvalues of discrete and continuous Laplace operators. Advances in Mathematics. 1996;**117**(2):165-178

[15] Wegner F. Bounds on the density of states in disordered systems. Zeitschrift Fur Physik B Condensed Matter. 1981;**44**:9-15

[16] Chulaevsky V. From fixed-energy localization analysis to dynamical localization: An elementary path. Journal of Statistical Physics. 2014;**154**:1391-1429

[17] Combes J-M, Thomas L. Asymptotic behaviour of eigenfunctions for multi-particle Schrödinger. Communications in Mathematical Physics. 1973;**34**:251-263

[18] Imbrie J. Multi-scale Jacobi method for Anderson localization. Communications in Mathematical Physics. 2016;**341**:491-521

[19] Brüning J, Geyler V, Pankrashkin K. Spectra of self-adjoint extensions and applications to solvable Schrödinger operators. Reviews in Mathematical Physics. 2008;**20**(20):1-70

Chapter 8

Math Words and Their Dendrograms

Francisco Casanova-del-Angel

Additional information is available at the end of the chapter

http://dx.doi.org/10.5772/intechopen.68531

Abstract

A hierarchical clustering algorithm based on graph theory is presented, which, from generation of a path from a given vertex, builds a math word and calculates clusterization under an index. This is possible due to modification of Tarry's algorithm, through exchange of path elements. The unidimensional clustering index applied to σ gives us what I have called Tarry's hierarchy. From the definition of net word, cycle, tree, tree word, and vertex, a theorem on the relationship between vertices, lines, and letters of a labyrinth is shown, which allows the generation of words and their dendrograms with the application of the Euclidean distance. Practical use of these concepts provides possibilities of connections in arrangements for telephone centrals and robotic arms' paths.

Keywords: dendrograms, Tarry, math words, connexe labyrinth, telephone central, robotics

1. Introduction

The first person to study the combinatory properties of schemes was Leonard Euler in 1736, who studied the network of the seven bridges in Königsberg, and he wrote in Berlin 1739 "*in Königsberg Pomeranie, they have a little island named Kneiphof, the river is divided by two and around the island seven bridges. You can arrange a network where you walk only one over in one time*". And Euler continues, "*this is possible for every one and impossible for others, but anyone has the certitude*" [1].

The modern theorem enunciated by Euler demonstrated that the necessity of the parity of the valence in every vertex is a theorem: *A connexe graphic is eulerienne if all vertex has degree pair.* Till date, the graphic theory has developed slowly but steadily. The principal contributors of this theory are Tarry [2, 3], Konig [4], Berge [5], Tutte [6], Bollobas [7], and Rosenstielh [8, 9, 10]. Recent works on graph theory are those of Bourdin et al. [11] and Gondran and Minoux [12].

The randomized Tarry algorithm is used in searching a graph of Urretabizkaya and Rodríguez 2004 [13] as they implement the Tarry algorithm for solving mazes of known structure. The graph theory is used in branches of mathematics like theory of groups, topology, theory of numbers, data analysis, and clusters. Among those who have contributed the most to the development of theory and models in telephone connections are Erlang, who implemented the well-known *Erlang Probability Density Function,* and his published works of 1917 and 1918 [14], and Elldin [15]. More recent contributors to this theorem are Ahuja et al. [16] and Mateus et al. [17].

As a starting point, let us see some graph theory definitions and concepts used in the generation of modification made to Tarry's algorithm, as well as the definition of applied distance, called minimum ultrametric distance in the application of Tarry's modified algorithm on telephone centrals arrangements. In the development of application, a theorem on the relationship between the maximum number of lines in a geometrical arrangement with n vertices and k letters is shown and demonstrated, which is made with the method known as mathematical induction.

2. Definitions and algorithms

Definition. One graphic is a couple $G = (X, U)$, where X is a finite set, of edges of G, each element of which is incident to two elements of another finite set V, named the set of vertices [5].

Definition. A labyrinth $\mathcal{L}$ is a finite set A non-empty and of pair cardinality, named set of words of $\mathcal{L}$ or alphabet, supplied of one involution l without an indeterminate point (i.e., without a tendency change) called by the prime operator:

$$l \in \mathrm{A} \text{ then } l' \in \mathrm{A}, l' \neq l \text{ and } (l')' = l \tag{1}$$

And one equivalence relation (named "with the same right than" where the classes are "the points" of labyrinth, the letters of same class have the same right as the point) indicated by the application of the letters which belongs to the letters in the set X of class:

$$\mathrm{d} : \mathrm{A} \rightarrow \mathrm{X} : \mathrm{d}(l) = \mathrm{d}(k) \tag{2}$$

By < the right of l is equal to right of k >, where < l has the same right as that of k > [9].

It is possible to speak of the left of letter too, making: g: $A \longrightarrow X$ with $g(l) = d(l') \; \forall \; l \in A$, as far as the labyrinth expressed by triad (A, i, d).

Definition. One labyrinth (A, i, d) is said to be orientated, if it exists one part A^+ of A such that:

$$l \in \mathrm{A}^+ \Leftrightarrow l' \notin \mathrm{A}^+ \quad \forall \; l \in \mathrm{A} \tag{3}$$

Definition. Given a labyrinth (A^+, i, d), σ is called a word of this labyrinth if it belongs to some of following classes:

http://dx.doi.org/10.5772/intechopen.68531

- $\Delta_\alpha \ \forall\, \alpha \in X$ is called empty words of $\mathcal{L}$
- l with $l \in A$ is called word letter of the $\mathcal{L}$, and
- $\sigma = l_{1},.., l_{r}, l_{r+1},\ldots, l_p$ with $l_r \in A \cup (\Delta_\alpha)\ \forall\, \alpha \in X$, we have $d(l_r) = g(l_{r+1})\ \forall\, r = 1,.., p\text{-}1$

Therefore, let X be the set of vertex, you can have lefts and rights applications in all words of the labyrinth $\mathcal{L}$ defined as d, g: $A \cup (\Delta_x)_{x\in X} \to X$ or d, g: $\mathcal{L} \to X$ with $g(\Delta_x) = g(l_1)$ and $d(\Delta_x) = d(l_p)$, i.e., $g(\Delta_x) = d(\Delta_x) = x$. And remember, σ is cyclical if $g(\sigma) = d(\sigma)$. To continue the definitions of tree, pure word, and neutral word, see Refs. [7, 18].

Definition. One tree A is a graph $G = (X, U)$ connected and noncyclical.

Let us see below how to materialize the above definitions. Given is the structure of a labyrinth oriented L: $A^+ \cup A^-$, **Figure 1**, where sets of letters A^+ and A^- are defined as $A^+ = \{a\, b\, e\, f\, i\, j\, c\, d\, g\, h\, k\, l\}$ and $A^- = \{a'\, b'\, e'\, f'\, i'\, j'\, c'\, d'\, g'\, h'\, k'\, l'\}$. In addition, the vertex set of L is $X = \{\alpha\, \beta\, \epsilon\, \gamma\, \lambda\, \delta\, \zeta\, \eta\, \xi\}$. It must be taken into account that L is the application of labyrinth and L is the labyrinth which, in this case, is the union of a finite set $A^+ \neq \phi$, with direction toward the right, and a finite set $A^- \neq \phi$, with direction toward the left, and X is the vertex set of the labyrinth.

A line of $\mathcal{L}$ would be $\{a\, a'\} \subset U_{\mathcal{L}}$, with $U_{\mathcal{L}}$ being the set of lines of $\mathcal{L}$. Some words of $\mathcal{L}$ with Δ_α being the empty word are $\{e\, f\, j\, i\}$, $\{a'\, b'\, f\, g\}$, $\{g\, d\, f'\, i'\, j'\}$, and $\{\Delta_\alpha e\}$. Cardinality of $\mathcal{L}$ is as follows:

$$\text{Card L} = \text{Card A} = \text{Card}(\text{A}^+ \cup \text{A}^-) = \text{Card A}^+ + \text{Card A}^- = 12 + 12 = 24 \tag{4}$$

A net word would be $\sigma = \{e\, f\, j\, i\, b\, a\, a'\, b'\, i'\, j'\, g\, h\, l\, k\, d\, c\, c'\, d'\, k'\, l'\, h'\, g'\, f'\, e'\}$, and right and left applications of σ are, respectively, as follows:

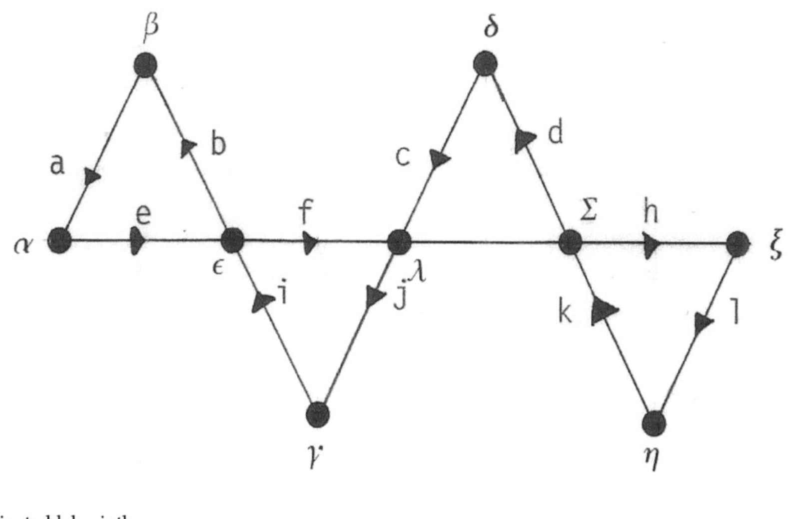

Figure 1. Oriented labyrinth.

$$d(\sigma) = \{e\,f\,g\,h\,l\,k\,d\,c\,j\,i\,b\,a\}\text{ and } d'(\sigma) = \{a'\,b'\,i'\,j'\,c'\,d'\,k'\,l'\,h'\,g'\,f'\,e'\} \tag{5}$$

It must be noticed that σ is cyclic: $d(\sigma) = g(\sigma)$ with $g(\sigma) = g(a) = a'$ and $d(\sigma) = d(e) = e$. We may also build a tree A_0, **Figure 2**, a tree net word and its dendrogramatic relationship. The hierarchical relationships that shall be seen are classifications with an aggregation criterion in accordance with the oriented path of the labyrinth under analysis.

It must be noticed that the word created based on the tree built is cyclic and its hierarchy does not exist, since it is nested (**Figure 3a**). Such nesting means that there is no aggregation criterion among the letters of the labyrinth alphabet (**Figure 3b**).

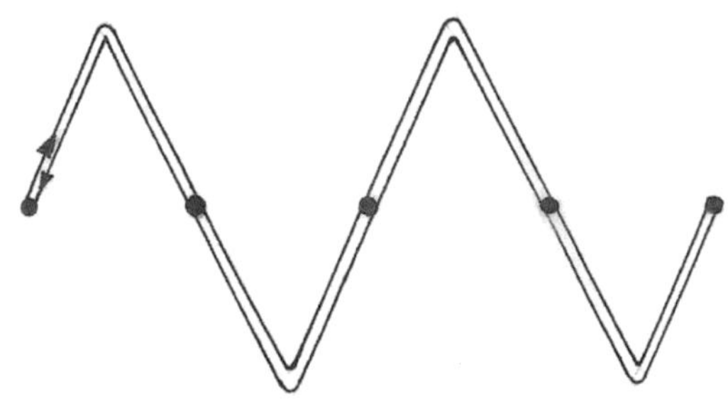

Figure 2. Tree A_0 of labyrinth oriented $\mathcal{L}$.

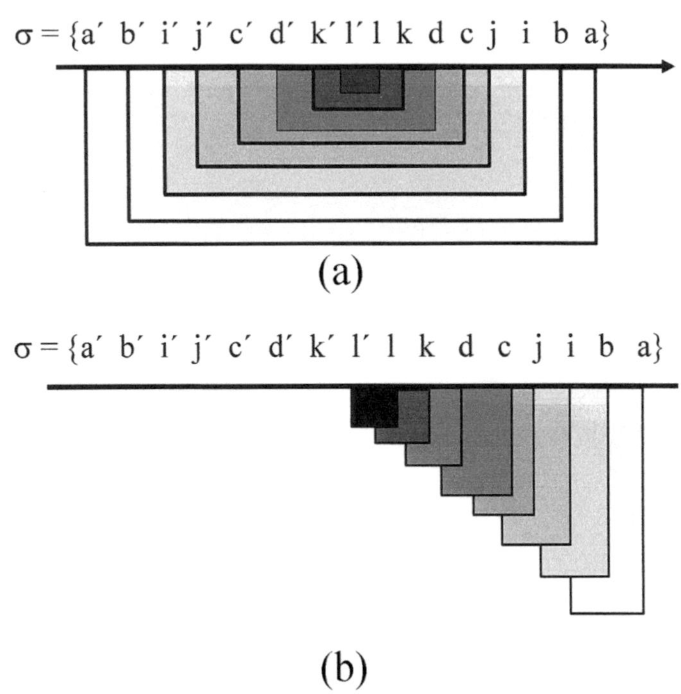

Figure 3. (a) Word created after tree A_0 (**Figure 2**), and (b) free aggregation criteria.

From labyrinth of **Figure 1**, we may build another tree and another tree net word, as shown in **Figures 4a**, **b**. The net word of tree A_1, **Figure 5a**, and his free aggregation criteria, **Figure 5b**, is as follows:

The new σ word is not completely cyclic, but even then, a hierarchy for such may not be built, since it is, mostly, nested.

Definition. If σ of $\mathcal{L} = (A, i, d)$, it is a pure word if inside σ exists an occurrence of each letter of A.

Definition. If $\sigma \in \mathcal{L}$ and $\sigma \approx \Delta_\alpha \; \forall \; \alpha \in X$, σ is called neutral word in α if the neutral element Δ_α is a neutral word in some α-particular.

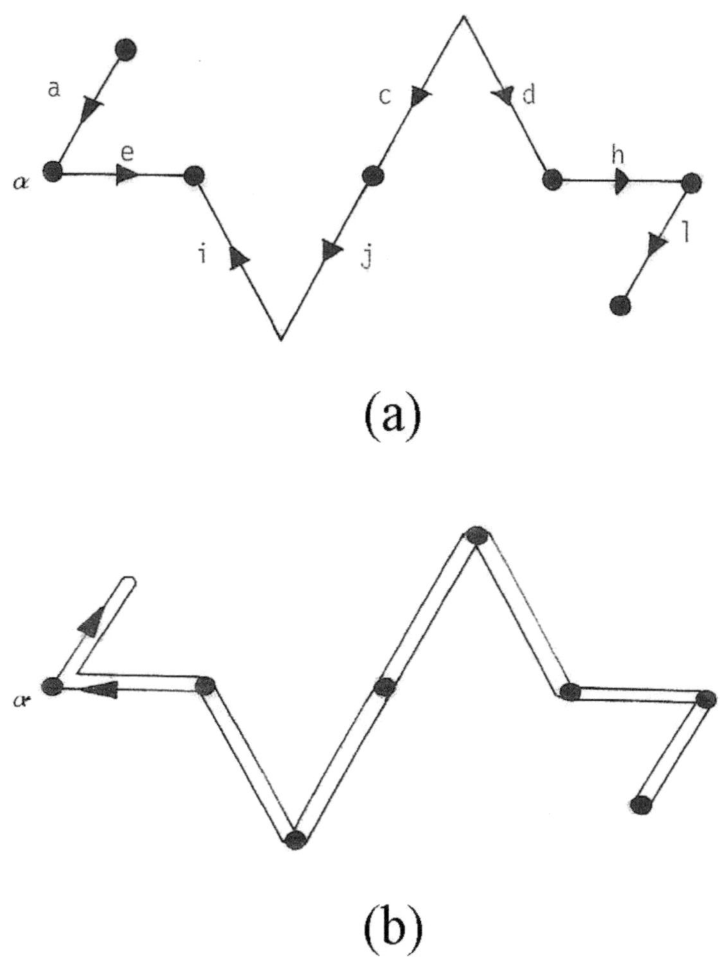

Figure 4. (a) Tree and (b) tree net word.

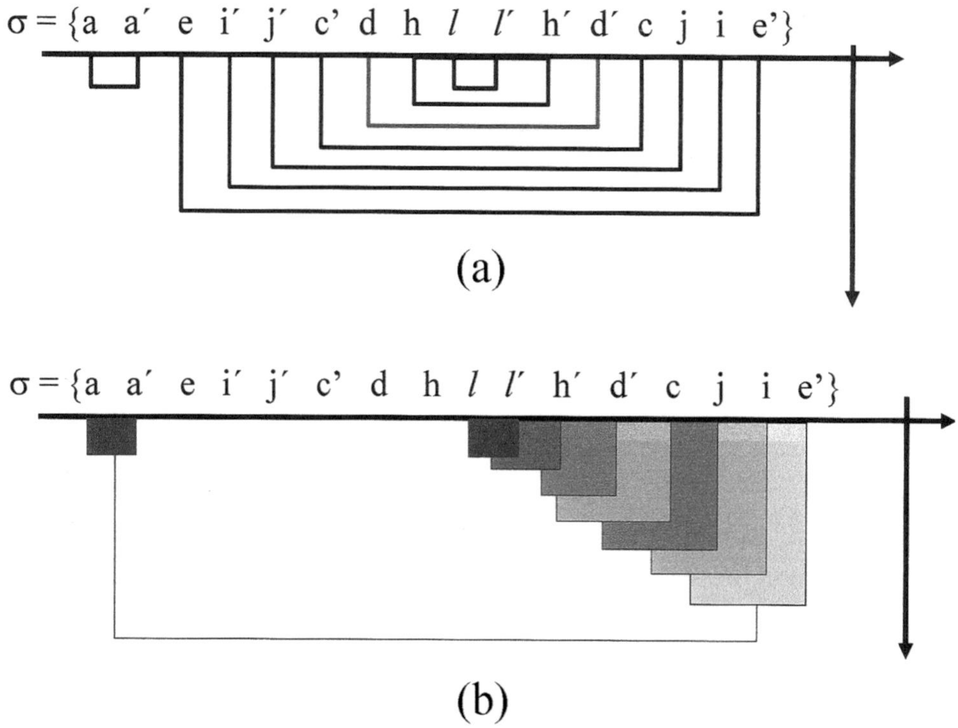

Figure 5. (a) The net word created after tree A_0 (**Figure 4a**) and (b) free aggregation criteria.

2.1. Algorithm of Tarry

Definition. Let $\mathcal{L}$ labyrinth be connected, we say Tarry's word of $\mathcal{L}$ all pure word $\Theta \in \mathcal{L}$ such that the entry tree V of Θ and the out tree W of Θ are opposites, i.e., $W = V'$.

Let L be the lexical of connected labyrinth $\mathcal{L} = (A, i, g)$, with $l_1 \in A$. Let $V(\sigma)$ be the entry tree of σ with σ a left factor of Tarry's word Θ to apply the following algorithm (**Figure 6**) [3]:

T.0 To put $\sigma \leftarrow l_1$

T.1 If $\sigma l \in \mathcal{L}$ with $l \notin \sigma$ and $l' \notin V(\sigma)$
to put $\sigma \leftarrow \sigma l$
but

T.2 If $\sigma l \in \mathcal{L}$ with $l \notin \sigma$ to put $\sigma \leftarrow \sigma l$
if
but

T.3 Stop. Write $\Theta = \sigma$
Where $\Theta = \sigma$ is called Tarry's word
Occasionally

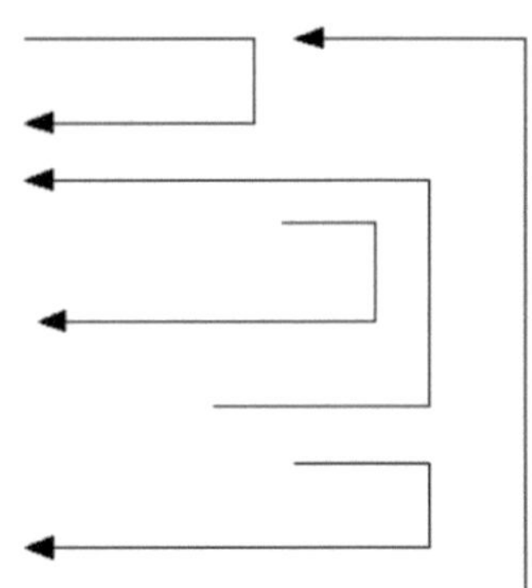

Figure 6. Graphic labyrinth of Tarry.

Note: The contrary of all entry letter is an out letter in $\Theta = \sigma$, cause Θ is a cyclical word, and all letter of A has the same right than the same left of Θ [19].

2.2. Attachment index

With the classification level of ways connected or paths of labyrinth $\mathcal{L}$, they begin with the first cycle or pleat of the Tarry's word. In this case, the index level is unit. When the Tarry's word σ is formed, you have to replace σ, writing always in first place the first obtained cycle.

Definition. Let two paths l_1 and $l_2 \in \mathcal{L}$ and let α all letters with the same left g(Θ) and β all letter with the same right d(Θ), the minimal distance d_M or the inferior ultra-metric minimal distance over a point $x_0 \in l_1$ and $x_1 \in l_2$ is $d_M(l_1, l_2) = \min\{d(x_0, x_1) \mid x_0 \in l_1 \;\&\; x_1 \in l_2\}$

Remember, the algorithm is applied after finding the Tarry's word and his cluster by couples or pleats. To finish the last pleat, begin the letter's arrangement in σ. You must begin with the first pleat.

2.3. Tarry's algorithm in pseudocode

In order to describe briefly the operative principle of modified Tarry's algorithm, its procedure is shown below in pseudocode, where, based on which, one may develop it in a programming language under conventional programming.

```
//Define the set of A+ letters with right ℒ//
  For i ←0 up to n with step 1 Make
     l1 ← A+
       i = i+1
  End for every li
//Define the set of A- letters with left ℒ//
  For i ←0 up to n with step 1 Make
     l'1 ← A+
       i = i+1
  End for every l'i
 //Define the set of vertices X = (x0, …., xn) of ℒ//
 For i ←0 up to n with step 1 Make
  x0 ← Xi
     In accordance with starting vertex x0 and right or left direction, Make
          σ ← li
       If σli ∈ ℒ with li ∉ σ and l'i ∉ V(σ) Make
          σ ← σli Make
          Θ = σ   //Means to build, letter by letter, Tarry's word//
       If not, Then change direction and Make
          σ ← σli+1
       If σli+1 ∈ ℒ with li+1 ∉ σ and l'i+1 ∉ V(σ) Make
          σ ← σli+1 Make
          Θ = σ  //Means to build, letter by letter, Tarry's word//
     End of right or left index of letters and increase their index
 End of vertex and increase its index
 Write
   Θ = σ    //Tarry's word//
 End of Tarry's word construction
```

//In Tarry's algorithm, there is the eventuality to go beyond this point to the first step if, by L's own nature, there is no other way to construction of Tarry's word, thus making a truncated Tarry's word//

End of Procedure

2.4. Application in basic telephone centrals arrangements

One of the great problems in telephony is to calculate telephone traffic increase, which allows designing telephone centrals and calculating the number of lines which shall satisfy the demand. Let us take the real case of **Figure 7** and apply the theory referred to above in order to obtain Tarry's trees, words, dendrograms, or clustering for optimal connection paths. If we take only the lowest part of **Figure 7** (telephone centrals and area centrals, **Figure 8**), the $\mathcal{L}$ labyrinth may be built:

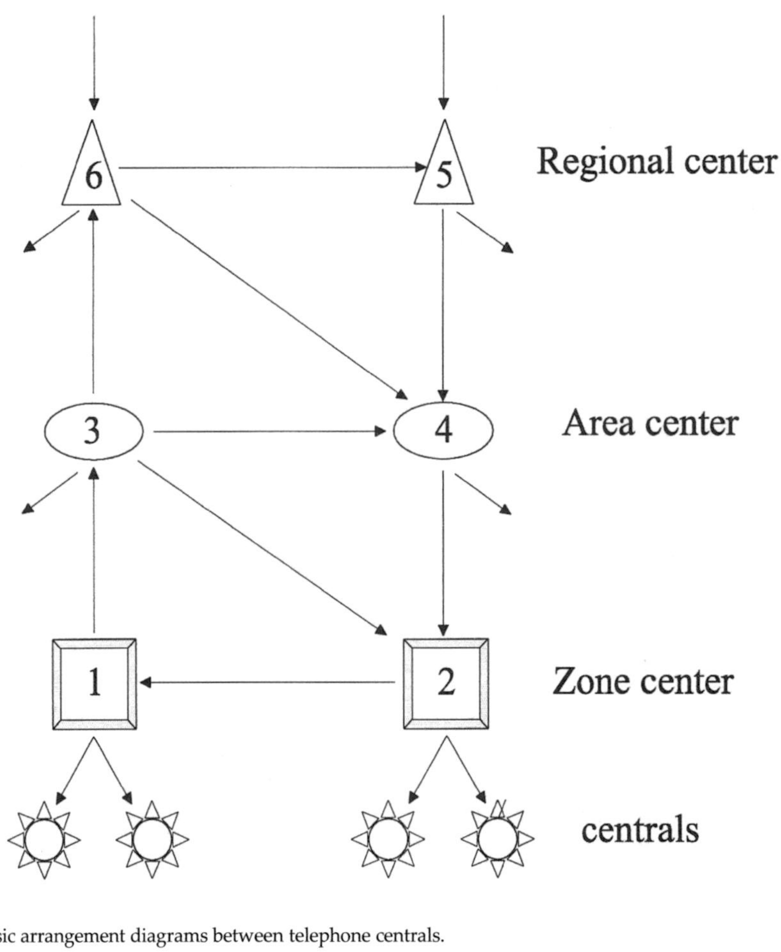

Figure 7. Basic arrangement diagrams between telephone centrals.

$$\mathcal{L} = \{a, b, d, e, a', c, b', c', d', e'\} \quad (6)$$

The number of trunks or telephone lines is $m = 5$, and the number of letters is $2m = 10$. If $x_0 \in X$ and $\{a, a'\} \subset A$, the net word of **Figure 8** is obtained. **Figure 9** is the net word of the $\mathcal{L}$ labyrinth:

Thus, seven trees related to vertex x_0 are obtained. To every tree corresponds one or several words (net, neutral, or Tarry's). Every tree thus built has 4 vertices, 5 lines and 10 letters. **Figure 10** shows all the trees related to x_0 vertex. It must be remembered that the same trees may be found in every vertex different from x_0.

Definition. A graphic $G = (X, U)$, such that $(x, y) \in U \Rightarrow (y, x) \in U$ is said to be symmetrical.

Based on the above definition, a telephone communications network is a symmetrical graphic, since every pair of terminals or adjacent vertices is linked in two directions.

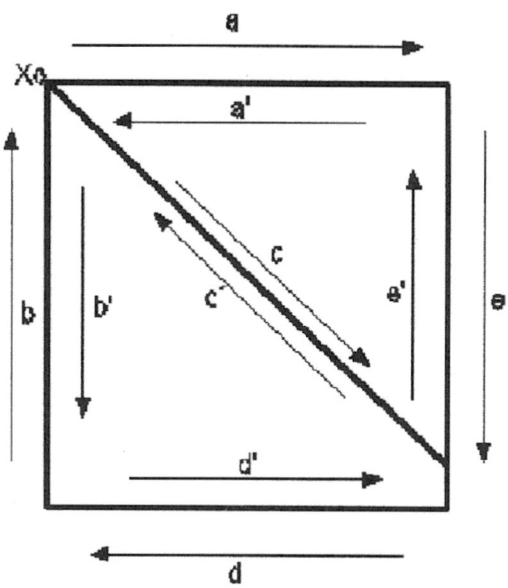

Figure 8. Tree diagrams of connections between telephone centrals and area centrals.

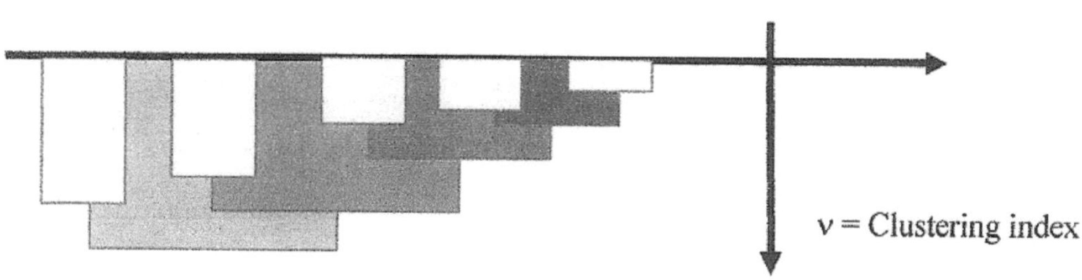

Figure 9. Net word of labyrinth.

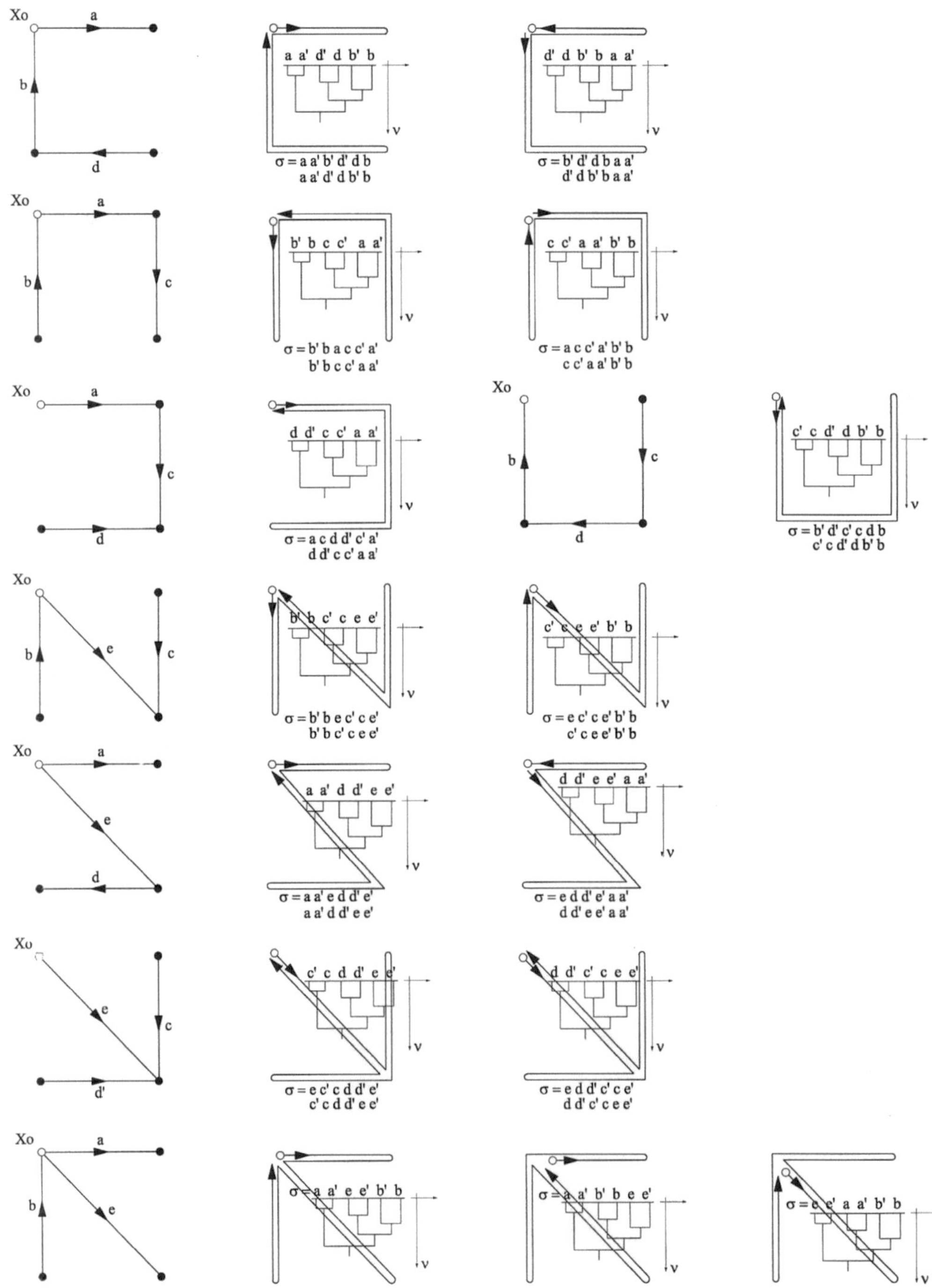

Figure 10. Trees, words, letters, and dendrograms of vertex x_0 in our basic arrangement between telephone centrals.

http://dx.doi.org/10.5772/intechopen.68531

Theorem. Let n be the number of vertices; m the maximum number of lines; and k the number of letters. If loops are not considered and we only study triangular geometrical arrangements, the relationship on the maximum number of m lines in a set of n vertices and k letters is:

$$n \leq m = \frac{(n-1)n}{2} \leq 2m = k \forall n > 2 \tag{7}$$

Demonstration. This algorithm shall be demonstrated through the induction mathematical concept (tested for one, supposed for n, and tested for n+1).

If $n = 1$, there is no labyrinth. If $n = 2$, then $m = \frac{(n-1)n}{2} = 1$ and $k = 2$. In this case, there is only one line in two directions (⇆). Let us suppose that $n = n$, then:

$$n \leq m = \frac{(n-1)n}{2} \leq 2m = k \forall \quad n > 2 \tag{8}$$

If we ask if this is valid for $n = n$+1. Then

$$\begin{aligned} n+1 \leq m &= \frac{(n+1-1)(n+1)}{2} \leq 2m = k \\ n+1 \leq m &= \frac{n(n+1)}{2} = \frac{(n^2+n)}{2} \leq 2m = \frac{2(n^2+n)}{2} = n^2+n = k \\ n+1 \leq m &= \frac{(n^2+n)}{2} \leq 2m = n^2+n = k \\ n \leq \left[\frac{(n^2+n)}{2}\right] &- 1 \leq m = n^2+n-1 = k \end{aligned} \tag{9}$$

that is, the number of vertices ≤ maximum number of lines ≤ number of letters, which complies with the theorem.

From our example:

$$\begin{aligned} &\text{if } n = 2 \quad \text{then } n = 2 \leq m = \frac{n(n-1)}{2} = \frac{2(2-1)}{2} = \frac{2}{2} = 1 \leq 2m = 2*1 = 2 = k \\ &\text{if } n = 3 \quad \text{then } n = 3 \leq m = \frac{n(n-1)}{2} = \frac{3(3-1)}{2} = \frac{6}{2} = 3 \leq 2m = 2*3 = 6 = k \\ &\text{if } n = 4 \quad \text{then } n = 4 \leq m = \frac{n(n-1)}{2} = \frac{4(4-1)}{2} = \frac{12}{2} = 6 \leq 2m = 2*6 = 12 = k \end{aligned} \tag{10}$$

When we apply the hierarchical distances geometrical theory [20] to build the tree or dendrogram of the labyrinth net word, **Figure 11**, hierarchical geometry shown is a starting sequence of isosceles triangles crowned by a sequence of scalene triangles. Partial hierarchies show a monotonous series, and in the following selection of classes to be added, there are always partial hierarchies corresponding to a primary telephonic connection. As for dendrograms of every word built based on vertex x_0, **Figure 12**, it may be seen that the hierarchy does not

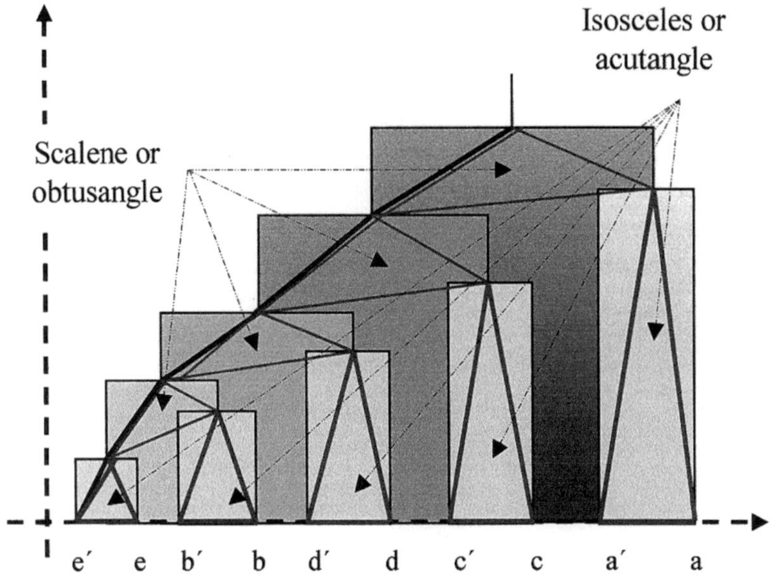

Figure 11. Dendrogram of the labyrinth's net word.

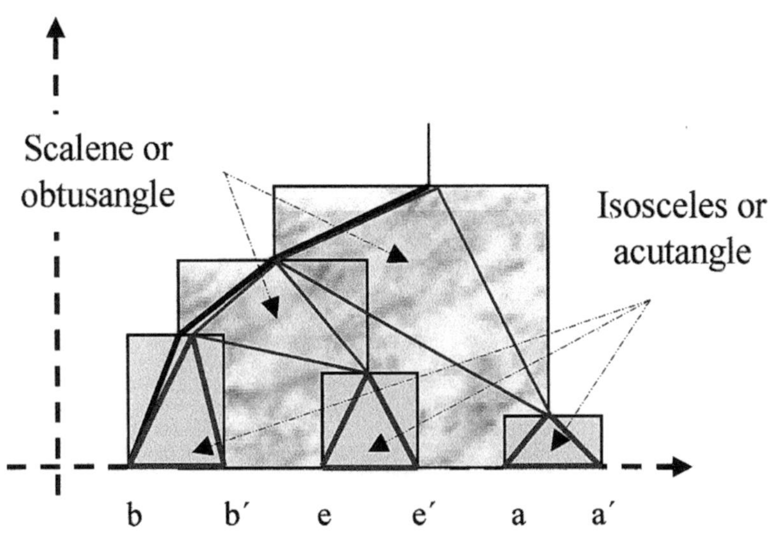

Figure 12. Hierarchical structure of tree, words, and dendrograms of vertex x_0.

change for each tree. Their geometry is identical for each different mathematical word obtained from the labyrinth.

Based on construction of dendrograms, the values of fractal dimension, its standard deviation, and que regression line equation have been obtained (**Table 1**). In the last column, the table shows to graphs: that of the fractal of the peak of the dendrogram and the geometrical

http://dx.doi.org/10.5772/intechopen.68531

structure of the hierarchy. Obtained values have been corroborated using Software Benoit 1.3 [21]. It must be considered that σ measures probability that an observation is at a certain distance from average observation, and it is valid if system under analysis is random.

Let us now talk about the polygonal case of the labyrinth's net word, as well as that representing all the trees, words, and dendrograms of vertex x_0, showing the sequence of variables incorporations into the hierarchical analysis developed. It must be considered that figures in **Table 1** are not at the same hierarchical scale. In the first one, case in which the dendrogram built is only for the labyrinth's net word, the polygonal sketches the trend of the final structure of each basic arrangement between telephone centrals, which is, in this case, the generator of fractal curves characterizing the dendrograms for each communication path between telephone centrals.

The segments of the generator are built linearly from the base of the class to the terminal point of a higher class, so that the sequence of such set of classes is conceptually defining part of the behavior of the net word's letters. The fractal's generator defines the characteristic of data under study, which in fact gives value to the construction of the hierarchical dendrograms [22].

The first four columns of **Table 1** register the names of the fractal dimension's estimation methods, with the value obtained for standard deviation and the regression equation obtained by such

Fractal of the labyrinth's net word

Estimation method	Fractal dimension	Standard deviation σ	Equation of the regression line
Box dimension	1.93693	0.0165441	$8.6E+05*X^{-1.94}$
Perimeter area	1.00533	0.1154578	$0.075*X^{1.99}$
Information dimension	1.96377	0.0012881	$8.76E+05*X^{-1.96}$
Mass dimension	1.9538	0.0073867	$3.95*X^{1.95}$
Ruler dimension	1.03249	0.000271	$4.57E+03*X^{-1.03}$

Fractals of trees, words, and dendrograms of vertex x_0

Estimation method	Fractal dimension	Standard deviation σ	Equation of the regression line
Box dimension	1.93695	0.0165482	$8.6E+05*X^{-1.94}$
Perimeter area	1.00532	0.1155108	$0.075*X^{1.99}$
Information dimension	1.96381	0.0012865	$8.75E+05*X^{-1.96}$
Mass dimension	1.95395	0.0074315	$3.95*X^{1.95}$
Ruler dimension	1.03249	0.000271	$4.57E+03*X^{-1.03}$

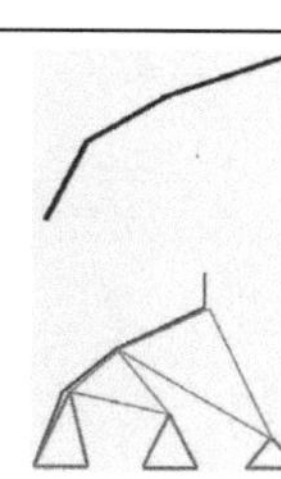

Table 1. Fractals of dendrogramatic labyrinth's math words.

method. Since it is a well-known theory, the description of such is not described in detail [23, 24, 25]. The last column, as described above, shows the fractal and its geometrical structure.

2.5. Application to industrial robots' paths

From the point of view of robotics, a robot is the technique applying informatics to design and use of equipment which, instead of people, carries out operations or tasks, generally, in industrial facilities. There is the problem of controlling kinematic and dynamic models. To define the movement of a robot implies controlling such in order that it may follow a pre-planned path. Therefore, the objective is to establish which are the paths to be followed by every articulation of the robot in time in order to achieve the objectives established, as well as it is required to comply with a series of physical restrictions imposed by actuators and the quality of the path, such as smoothness and precision.

In order to make an animation of a rotational robot with x-degrees of freedom, a simple planning of the path to be followed by the robot must be made, taking into account the actuators, so that the movement of the robot may be smooth and coordinated. The types of paths of industrial robots currently are as follows: point-to-point path, coordinated or isochrone path, and continuous paths. For 3D paths, related to 2D paths, the versatility of inverse kinematics and dimensions of existing algorithms are different.

Let us consider now an industrial robot making a complex set of paths in three dimensions to assemble a device, as shown in **Figure 13**. Here, the oriented path set for the mechanic arm and the points, in our case *vertices*, where such shall implant an element or component of the system being assembled are shown. Treatment of the path under the theory shown is as follows:

It must be said that through desired paths generated by a specific function, it must be segmented into enough points to describe follow-up on the path made by the robot manipulator. Thus, high accuracy may be achieved; however, the search process extends the convergence

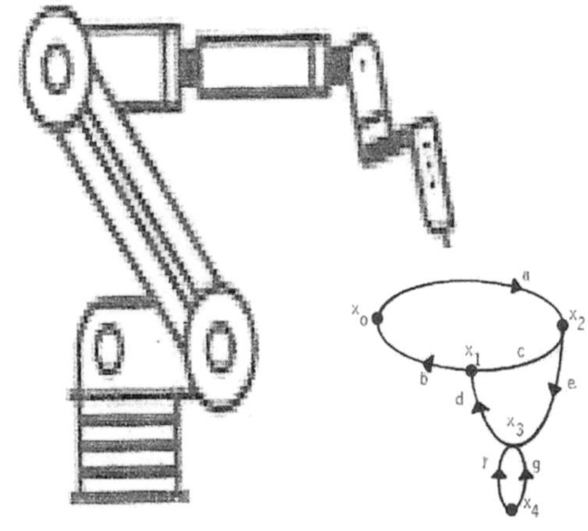

Figure 13. Robot with ellipsoidal-oriented paths.

http://dx.doi.org/10.5772/intechopen.68531

time of algorithms. This is also true for traditional numerical solutions, such as Newton-Raphson's, since it has a small differential step.

The sets of letters are: direction toward the right $A^+ = \{a\ b\ c\ d\ e\ f\ g\}$, direction toward the left $A^- = \{a'\ b'\ c'\ d'\ e'\ f'\ g'\}$, and that of vertices of $\mathcal{L}$ is $X = \{x_0,\ x_1,\ x_2,\ x_3,\ x_4\}$. **Figure 14** shows the non-neural Tarry's labyrinth word diagram. Therefore, a Tarry's word to see the circular permutations of letters and vertices is:

Word Θ built is, evidently, nonneural (**Figure 15**). Its diagram is shown in **Figure 16a**, but almost always another word may be built, **Figure 16b**, which is hierarchical:

Word Θ built is neutral but not Tarry's, since its folding into e does not allow it to be so. Its diagram is shown in **Figure 17**.

In this example, the informatics algorithms defining the paths of a robotic arm are related to the work cycles and co-cycles of their industrial task or partial graphics of the paths. This duality between cycles and co-cycles is extended to the concept of tree. **Figure 18** shows the graphic representation of entrance-exit or right-left trees. Based on labyrinth oriented $\mathcal{L}$,

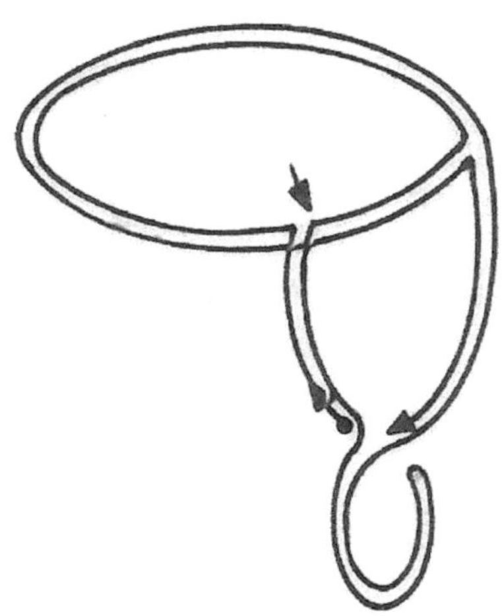

Figure 14. Non-neutral Tarry's labyrinth word diagram.

$$\Theta = \{d\ b\ a\ c'\ d'\ f'\ g\ g'\ f\ e'\ a'\ b'\ c\ e\}$$

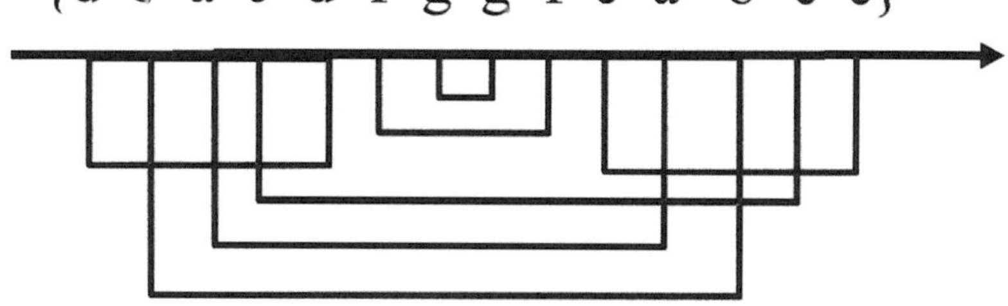

Figure 15. Word Θ non-neural.

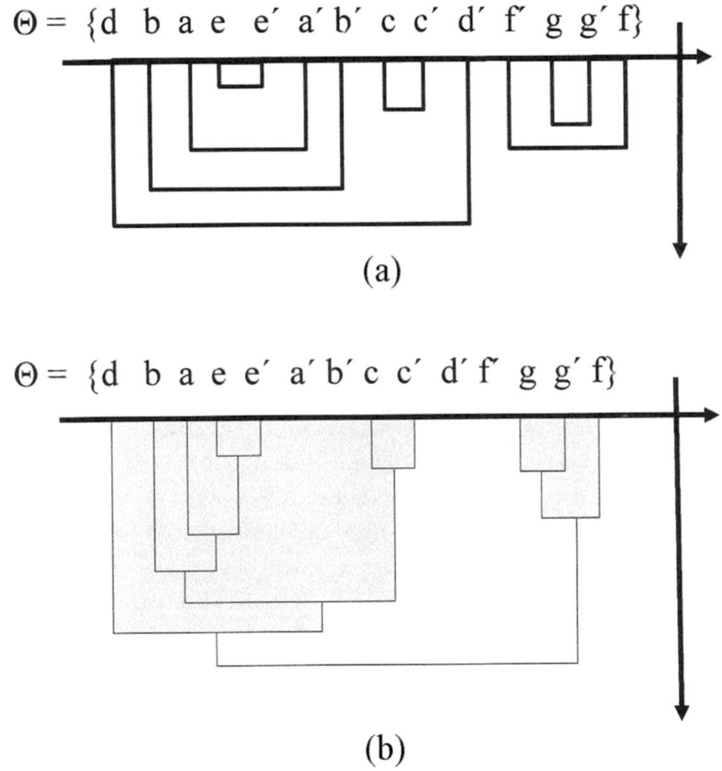

Figure 16. (a) Non-neural word of labyrinth $\mathcal{L}$ (**Figure 13**) and (b) neural non-Tarry's word.

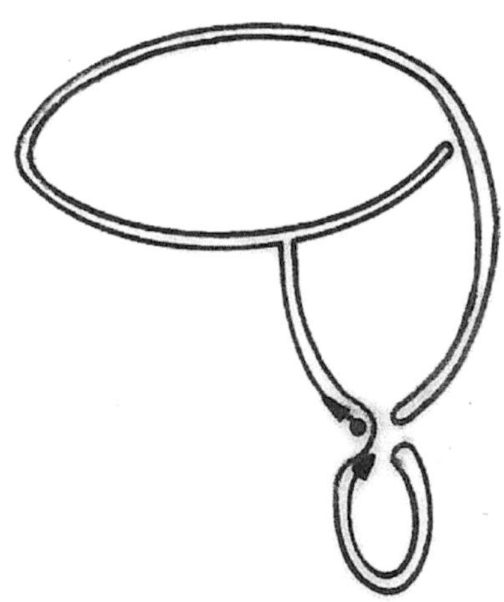

Figure 17. Neutral non-Tarry's Θ word diagram.

Figure 13, let us build a new μ Tarry's word, as well as its hierarchical relationship (**Figure 19**). **Figure 20** shows its diagram.

Definition of cocycle. A cocycle of vertex $x \in G$ is the addition of every $w(x)$ edge where an end is vertex x.

http://dx.doi.org/10.5772/intechopen.68531

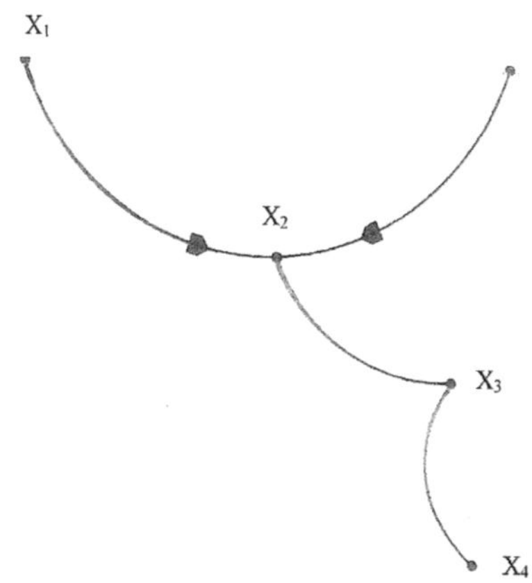

Figure 18. Θ's entrance and exit trees.

Based on labyrinth-oriented $\mathcal{L}$ described in **Figure 13**, a new ξ Tarry's word and its hierarchical relationship are built, which diagram is shown in **Figure 21**.

Inside an $\mathcal{L}$ labyrinth, it is possible to make permutations of letters (it had already been done here, but it had not been explained), as well as of lines, and vertices, respectively. If the labyrinth is connected, it is possible to make permutations of elements already mentioned, but in a circular manner.

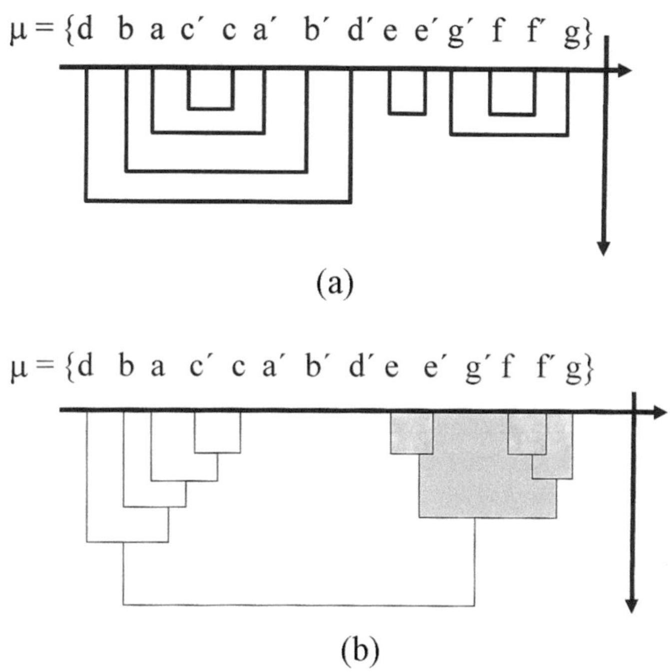

Figure 19. (a) μ Tarry's word form (**Figure 17**) and (b) its hierarchical relationship.

Figure 20. μ Tarry's word diagram.

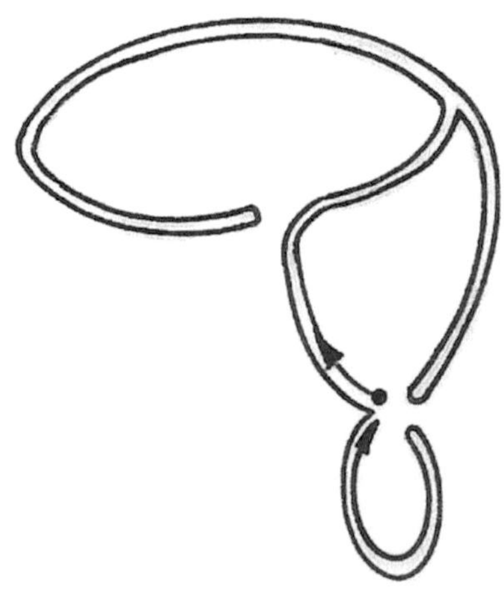

Figure 21. ξ Tarry's word diagram.

Definition. A circular transformation in a connected labyrinth is a substitution of the order of letters, lines, or points with another order, without negatively affecting the nature of such labyrinth.

Let $k \geq 0$ and Φ: $A \to U$ the application that relates every letter to the corresponding line or arch in labyrinth $\mathcal{L}$, where U is the family of arcs. Therefore, it is said that (i) two letters m_1 and m_2 are k-adjacent if there is a word of $\mathcal{L}$ with less than $k + 2$ occurrences of letters, where the first word is m_1 and the second word is m_2. (ii) The sets of words A_1 and A_2 are k-adjacent if there are two letters k-adjacent m_1 and m_2, such that $\Phi(m_1) = A_1$ and $\Phi(m_2) = A_2$ and (iii) two vertices x_1 and x_2 are k-adjacent if there are two letters, also k-adjacent, m_1 and m_2, such that $x_1 = g(m_1)$ and $x_2 = g(m_2)$.

Based on the above, let us now keep only the odd rank letters of word ξ

$$\xi = \{d \quad e \quad a' \quad b \quad c' \quad f' \quad g'\} = p_1 \tag{11}$$

Word ξ or p_0 is a circular permutation of letters with 0-adjacenty, that is, $k = 0$. ξ in p_1 is a circular permutation of lines with 1-adjacency (**Figure 22**). Using the entrance tree, **Figure 18**, the ξ word of the corresponding entrance tree is shown in **Figure 23a** and its hierarchical relationship in **Figure 23b**.

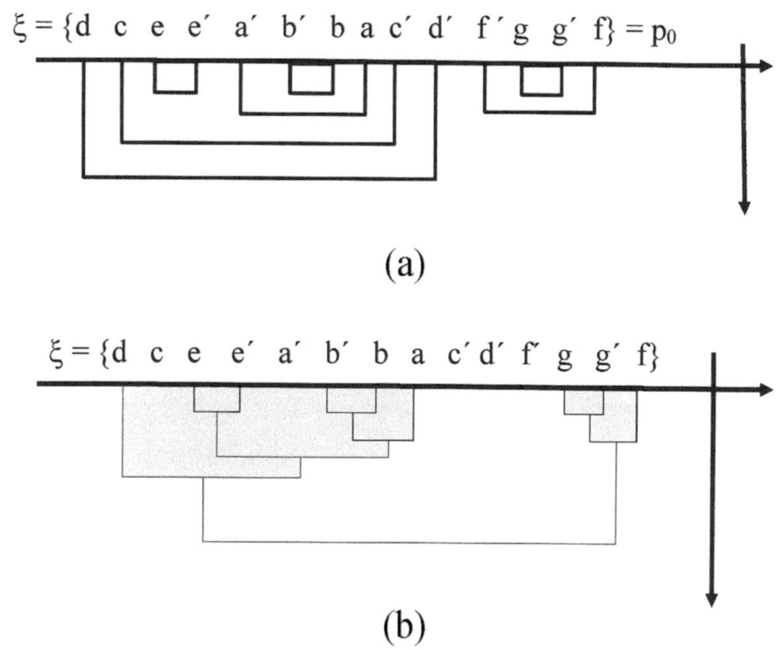

Figure 22. (a) New ξ Tarry's word form (**Figure 21**) and (b) its hierarchical relationship.

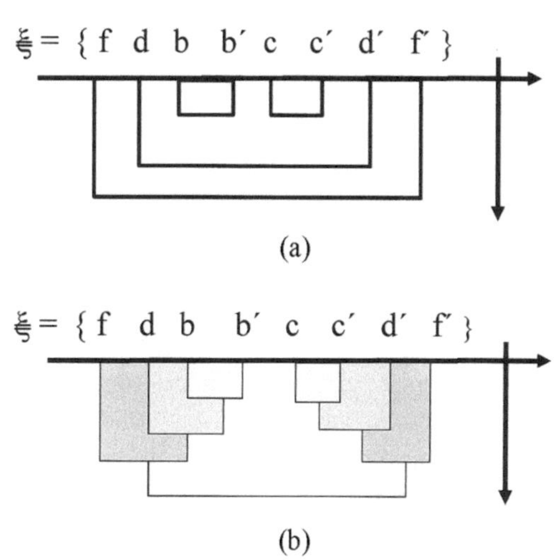

Figure 23. (a) The ξ word of the corresponding entrance tree and (b) its hierarchical relationship.

Keeping from ξ the odd rank letters and relating each of them to a vertex, in accordance with the rule, we obtain:

$$p_2 = \begin{cases} \quad f \quad\quad b \quad\quad c \quad\quad d' \\ x_4 \quad\quad x_3 \quad\quad x_0 \quad\quad x_2 \quad\quad x_3 \end{cases} \tag{12}$$

which makes p_2 a circular permutation of dots with 2-adjacency.

3. Conclusions

It is known that all expositions need a complete solution, with links' help, which suggests the path, as horizontally as and vertically, from algorithm's solution that insides in a formal language able to analyze the system to rebuild the principal structure. The modification in the algorithm of Tarry through exchange of elements allows a better arrangement or letters couple assembly to permit the mathematical word building. The unidimensional attachment index is applied to σ giving the Tarry's hierarchical.

There is no published clustering algorithm (whether hierarchical or not), based on graph theory, which, from generation of a path starting on a given vertex, builds a math word and calculates clustering under an index. This has been possible by modification to Tarry's algorithm, through exchange of elements, which has allowed a better arrangement of letters coupling and the construction of a math word. The unidimensional clustering index applied to σ gives what I call Tarry's hierarchy. The path planning problem may be approached with classical approaches, such as the potential fields, or more modern approaches, such as Fuzzy techniques [26], or the one shown here, through Tarry's modified algorithm, which takes into account the work cycles and co-cycles in the orientation of the labyrinth of paths defined to carry out a particular task. In addition, the set of defined paths has an order and hierarchy, for which reason, it is revealing the building of dendrogramatic trees based on net words built, which allows to analyze if work paths defined for the robotic arm are optimal.

Methodology for fractal analysis of geometric discontinuity or polygon created by unions of midpoints of vertices, nodes, or peaks of dendrograms' terminal classes is based on the following: (i) the outline of the dendrogram under study is drawn, linking the midpoints of the vertices, nodes, or peaks of the terminal classes, verifying the sectioned behavior of the polygon; (ii) any mirror behavior in any section of the curve is identified; (iii) the curve showing the direction of its fractal propagation is isolated and a straight line is drawn from the starting point to the ending point, calculating, regarding the ordinates axis, its propagation angle; (iv) the scale factor of the fractal curve is calculated; (v) the curve generator is reproduced to scale; (vi) meshing is created for calculation of the fractal dimension of the polygon; (vii) the fractal dimension for every different meshing and rotation is calculated, graphing the results in accordance with each estimation methods theoretically accepted; and (viii) the self-similarity property of the polygon is verified.

http://dx.doi.org/10.5772/intechopen.68531

Author details

Francisco Casanova-del-Angel

Address all correspondence to: fcasanova49@prodigy.net.mx

National Polytechnic Institute, Mexico City, Mexico

References

[1] Euler L. Solutio Problematis ad geometriam situs pertinentis. Mémoire de l'Academie des 3 Sciences de Berlin; 1739. Originally published in *Commentarii academiae scientiarum Petropolitanae* 8, 1741, pp. 128–140. *Opera Omnia*: Series 1, Volume 7, pp. 1-10. Reprinted in *Comment. acad. sc. Petrop.* 8, ed. nova, Bononiae 1752, pp. 116-126 + 1 diagram [53a]. A handwritten French translation of this treatise can be found in the library of the observatory in Uccle, near Brussels

[2] Tarry G. Parcours d'un labyrinthe rentrant. Assoc. Franç. Pour l'Avanc. Des Sciences; 1886. pp. 49-53

[3] Tarry G. Le problème des labyrinths. Nouvelles annales de Mathematiques. Vol. XIV; 1895

[4] Konig D. Theorie der Endlichen und Unendlichen Graphen: Kombinatorische Topologie der Streckenkomplexe. Leipzig: Akad. Verlag

[5] Berge, C. Graphes et hypergraphes. Monographies Universitaires de Mathématiques n° 37, Paris, Dunod, 1970. English translation: «Graphs and hypergraphs», vol. 6, Amsterdam-London, North-Holland Publishing Co,; New York, American Elsevier Publishing Co. Inc., 1973. Second revised version: Paris-Brussels-Montréal, Dunod, Coll. Dunod-Université, série Violette, n° 604, 1973. English translation: vol. 6, Amsterdam-London, North-Holland Publishing Co.; New York, American Elsevier Publishing Co, Inc., 1976

[6] Tutte WT. On Hamiltonian circuits. Colloquio Internazionale sulle. Teorie Combinatoire. Atti Convegni Lincei 17. Accad Naz. Lincei. Roma I, pp. 193-199

[7] Bollobás B, editor. Advances in Graph Theory. Amsterdam: North-Holland; 1978

[8] Rosenstiehl P. Existence d'automates finis capables de s'accorder bien qu'arbitrairement connectés et nombreux. I.C.C. Bulletin. 1966;**5**(4): 245-261

[9] Rosenstiehl P. Labyrinthologie Mathematique. Mathematique et Sciences Humaines, 9e année, Num.1971;**33**:5–32

[10] Rosenstiehl P. Les graphes d'entrelacement d'un graphe. Coll. Inter. CNRS n°. 260, Problèmes Combinatoires et Théorie des Graphes, Orsay. Ed. du CNRS. 1976. pp. 359-362

[11] Bourdin H, Ferreira A, and Marcus K. A performance comparison between graph and hipergraph topologies for passive star WDM light wave networks. Computer Networks and ISDN Systems. 1998;**8**(30):805819

[12] Gondran M, Minoux M. Graphes et algorithmes, Ed. Eyrolles, coll. « Dir. Ét. & Rech. EDF », (1979), 622 p. EAN13: 9782212015713

[13] Urretabizkaya R, Rodríguez O. Análisis e implementación de algoritmos para la solución de laberintos de estructura conocida. México: Universidad Autónoma de Querétaro; 2004

[14] Erlang AK. Solutions of some problems in the theory of probabilities of significance in automatic telephone exchanges. The Post Office Electrical Engineers' Journal. 1917;**10**(4): 189-197

[15] Elldin A. Automatic Telephone Exchanges with Crossbar Switches: Switch Calculations: General Survey. Publisher Ericsson. 1961. 37 pages

[16] Ahuja RV, Magnanti TL, Orlin JB. Network Flows: Theory, Algorithms and Applications. Englewood Cliffs, NJ: Prentice Hall; 1993

[17] Mateus GR Cruz, F. R. B. and Luna HPL (1994). An algorithm for hierarchical network design. Locations Science. 1994;**2**:149-161

[18] Golumbic MCh. Algorithmic Graph Theory and Perfect Graphs. USA: Academic Press; 1980

[19] Casanova-del-Angel F. Introducción a la teoría de gráficas, a los laberintos y a sus palabras. Boletín de Graduados e Investigación, vol. I, núm. 3, Mexico: Instituto Politécnico Nacional. pp. 23-51

[20] Casanova-del-Angel, F. Structure géométrique des distances hiérarchiques. Revue MODULAD de data mining, Statistique et Analyse des données. Année 2010, numéro 41 (janvier 2010), ISSN :17697387. France: Institut National de Recherche en Informatique et en Automatique; 2010. pp. 27-37

[21] Benoit 1.3. 2008. www.truesoft-international.com/benoit.html

[22] Casanova-del-Angel F. Structure fractale d'un dendrogramme. Revue MODULAD de data mining, statistique et analyse des données. Année 2011, numéro 43 (février 2011), ISSN:17697387. France: Institut National de Recherche en Informatique et en Automatique. pp. 44-55

[23] Jian L, Qian D, Caixin S.. An improved box-counting method for image fractal dimension estimation. Pattern Recognition. 2009;**42**(11):2460-2469. DOI: 10.1016/j.patcog.2009.03.001

[24] Camastra F. Data dimensionality estimation methods: A survey. Pattern Recognition. 2003;**36**:2945-2954. DOI: 10.1016/s0031-3203(03)00176-6 (2003)

[25] Theiler J. Estimating fractal dimension. Journal of the Optical Society of America A. 1990;**7**:1055-1073

[26] Yen J, Langari R. Fuzzy Logic intelligence control and information. New Jersey: Prentice Hall; 1999. pp. 3519-362